PHILOSOPHY OF NATURE

PHILOSOPHY OF NATURE

JACQUES MARITAIN

To which is added

**MARITAIN'S PHILOSOPHY
OF THE SCIENCES**

By YVES R. SIMON

PHILOSOPHICAL LIBRARY
NEW YORK

TRANSLATED FROM THE FRENCH
BY IMELDA C. BYRNE

PRINTED IN THE UNITED STATES OF AMERICA

Author's Preface

In the preface to our *Seven Lectures on Being* [1] we stated our intention of publishing a series of lectures on the Philosophy of Nature. The lectures which make up the present work are the first of this series. We hope to follow them by lectures on Matter and Form and on the Living Organism.

Here, as in the *Lectures on Being*, we must ask the reader to excuse the familiarity of style, the digressions and repetitions characteristic of lectures which have been taken down in short-hand. We hope that these inconveniences will be offset by the livelier and more extensive presentation of the subject-matter which such a style makes possible.

J. M.

Translator's Preface

I want to thank Mr. Maritain for reading the manuscript before printing and for the corrections he made. My thanks go, too, to Rev. Gerald B. Phelan, University of Notre Dame, and to my friends at the Pontifical Institute of Mediaeval Studies, University of Toronto, for their suggestions and helpfulness.

I am also grateful to Dr. Donald Gallagher of Marquette University, who suggested that the Bibliography and Professor Simon's article be added, and to Professor Yves Simon and Sheed and Ward for permission to reprint this article, which first appeared in the Maritain Volume of *The Thomist* (1943).

IMELDA CHOQUETTE BYRNE

Table of Contents

	PAGE
AUTHOR'S PREFACE	v
TRANSLATOR'S PREFACE	vii

CHAPTER I

THE ANCIENT PHILOSOPHERS' CONCEPTION OF THE
PHILOSOPHY OF NATURE; ITS DIFFICULTIES ... 1

Section 1. Greek and Mediaeval Philosophy ... 4

Heraclitus and Plato, 4. Aristotle, 7. The Orders of
Abstractive Visualization, 12. Metaphysics, the Philosophy of Nature and the Natural Sciences, 31.

Section 2. The Galileo-Cartesian Revolution ... 36

The Intermediary Sciences, 36. A Tragic Misunderstanding, 41.

CHAPTER II

THE POSITIVISTIC CONCEPTION OF SCIENCE, AND ITS
DIFFICULTIES ... 45

Section 1. The Positivistic Conception of Science ... 45

The Genesis of the Positivistic Conception, 45. The
Advent of Empiriological Thought and the Con-
cept of Science, 49. The Advent of Empiriological
Thought and Metaphysics, 55.

Section 2. Modern Reactions Against the Positivistic Conception of Science ... 60

Pierre Duhem, 60. Emile Meyerson and French Epistemology, 62. German Phenomenology, 71.

CHAPTER III

PAGE

THOMISTIC POSITIONS ON THE PHILOSOPHY OF NATURE · 73

Section 1. Necessity of the Philosophy of Nature · 73

Empiriological and Ontological Analysis, 73. The Philosophy of Nature Differs Specifically from the Natural Sciences, 89. The Philosophy of Nature and the Natural Sciences Are Mutually Complementary, 93. Answer to a Difficulty, 98. The Subordination of the Empiriological Realm to Mathematics or to the Philosophy of Nature, 102. Applications to Biology, 114.

Section 2. Definition of the Philosophy of Nature · 118

The Philosophy of Nature and Metaphysics, 118. The Philosophy of Nature and the Sciences, 123. Formal Objects and Formal Perspectives, 125. The Philosophy of Nature and the Empirioschematic Sciences, 135. The Philosophy of Nature and the Empiriometrical Sciences, 138. Proposed Definition of the Philosophy of Nature, 139. The Philosophy of Nature and Facts, 140. The Contemporary Renaissance of the Philosophy of Nature, 151.

CHAPTER IV

MARITAIN'S PHILOSOPHY OF THE SCIENCES—By Yves R. Simon · 157

SELECTED BIBLIOGRAPHY · 183

FOOTNOTES · 191

INDEX TO PROPER NAMES · 197

I

The Ancient Philosophers' Conception of the Philosophy of Nature; Its Difficulties

1. The philosophy of nature is caught between two opposed dangers: that of being absorbed by the experimental sciences which claim that knowledge of the sensible world, of the universe of nature, belongs to them alone, or that of being eclipsed by metaphysics. For many modern philosophers, following a tradition that goes back to Wolff, would like to include cosmology in the realm of metaphysics. Thus, be it absorbed by the sciences or eclipsed by metaphysics, it is very difficult for the philosophy of nature to defend its own existence.

During these lectures I hope that we shall see clearly the capital importance of these questions about the philosophy of nature, its 'autonomy' (as Driesch would say), its specificity as knowledge, its relations with the sciences on one hand and with metaphysics on the other. I have already treated somewhat of these matters in the *Degrees of Knowledge*,[1] but now I should like to take up the question again as a whole and with more precision and depth.

1

2. The dispute between philosophy and science is particularly keen with respect to a central problem: that of the philosophy of nature. Should a philosophy of nature which is at once distinct from metaphysics and from the particular sciences exist? What are its characteristic traits, its nature and definition, its spirit? These are not easy questions to answer for they come to us fraught with historical implications and complications. Is not the philosophy of nature that which Aristotle called Physics? For the ancients did not this 'Physics' include the whole domain of the natural sciences? Did not the downfall of the Aristotelian explanations of natural phenomena entail the downfall of Aristotelian physics as a whole and thus of the philosophy of nature? And therefore for us moderns what else should replace Physics in the Aristotelian sense than Physics itself,—but Physics in the sense of Einstein, Planck and Louis de Broglie or, more generally, in the sense of the whole ensemble of the sciences of natural phenomena,—what the modern world calls Science? Such are the connections and associations bound up with the theoretical questions with which we must deal.

These are not easy questions; they are fundamental. Let us not hesitate to say that, for wisdom, they are of first-class importance. The problem of the philosophy of nature must not be neglected. The philosophy of nature is the humblest, the nearest to the senses, the most imperfect of the speculative wisdoms; in the pure and simple sense of the word it is not even wisdom. It is wisdom only in the order of mutable, corruptible things. But that is precisely the order which is most proportioned to our thinking nature. It is the first wisdom we come to in the progressive and ascensional movement of our reason. Which is why it has so much importance for us: precisely because it is at the lowest rung of the φιλία τῆς σοφίας.

Whereby can the real enter into us? There are for us but two sources by which it may do so, the one natural, the other supernatural: the senses and the Spirit of God. With respect to the lights that come to us from above, not metaphysics but the highest, wholly spiritual wisdom is first, the wisdom of the saints. It is by this wisdom that we are open to the influence of those heavenly lights and that something enters into us through a special gift of grace. With respect to the lights that come to us from the material world, metaphysics is again not first. Here an inferior wisdom linked to sensory perception and strictly dependent on experience is first. For it is by the senses that we are open to things and that something enters into us in accordance with our natural mode of knowing.

Metaphysics stands between these two wisdoms: it is not directly intuitive of divine things, as the Platonists would have had it: metaphysical intuition crowns the process of visualization or abstraction which starts from the sensible. In itself it is formally independent of the philosophy of nature, being a superior and regulative knowledge. But materially and as to us, it presupposes

the philosophy of nature: not in its completed state, no
doubt, but at least in its first positions.

SECTION 1 · GREEK AND MEDIAEVAL PHILOSOPHY

3. How can we best describe man's first efforts to
speculate about nature, such for example as the history
of the Pre-Socratics gives witness to?

First of all, I would like to point out the major logical
articulations which we must keep in mind here.

Heraclitus and Plato

4. The intellect, as we know, is made for being: it
seeks it and in seeking being which is its connatural object,
well, it comes upon the sensible flux of the singular, of
the changing singular, and naturally it is disappointed.
It seeks being, it finds becoming, becoming which it
cannot grasp. So greatly is it disappointed that the intellect is tempted to make this world of becoming and
of the sensible singular consist of one great deception:
the *maya* of Hindu philosophy. I believe that the movement of the mind which I am trying to describe to you
here is truly the natural movement of the human intellect as it occurred in the first speculations of India and
Greece.

It is not at all surprising that this intellect should have
been discouraged in face of the flux of becoming wherein
it does not find the object for which it is made: being

with its intelligible necessities and stability. The great prophet of this intellectual discouragement was Heraclitus who asserted, as you know, that we cannot bathe in the same river twice, and held the knowledge of nature to be impossible. No doubt Heraclitus had a hidden and more or less mythical metaphysics of his own, but what is most important in the exposition of his thought and what struck him above all else, is the scandal in which the principle of contradiction is involved because of the fact of becoming. Instead of affirming this principle and denying becoming, as Parmenides did, Heraclitus tends to affirm the identity of contradictories in order to safeguard the reality of becoming. But by this very fact, becoming escapes the grasp of our intellect.

Plato is very near Heraclitus from the point of view we are talking about here, that is from the point of view of the knowledge of nature. He too sought being and found sensible flux and therefore he too, discouraged by this flux, declared that the world of sensible nature can only be the object of opinion, δόχα, not of science. When the mind's eye falls upon the flux of the sensible it must immediately turn back, away from it, to true science which is strong, solid, unshakable. This it does by contemplating the intelligible types separate from these sensible things which are caught in the flux of movement and change. What is to be contemplated by the metaphysical eye is the world of Platonic ideas: objects which are not only intelligible objects,—essences,—but which are considered under the logical conditions peculiar to ideas in our mind; in scholastic language, under *conditions of reason*, existing only within the mind. It is only in our mind that the universal enjoys the positive unity which is proper to it; it is in our mind that it is separate from things. Plato confers these two characteristics: positive unity and separation from things (characteristics which belong to the universal object of thought as it is in our mind), upon the eternal objects contemplated by the metaphysician.[2]

There is here a contamination of the real by the logical which fully explains why Plato applies the word *idea* to the highest realities. These ideas are, for example, man in himself, tree in itself, etc., and finally at the apex, the idea of the Good which is contemplated by simple intellectual vision, *noésis*, whereas the multiplicity of ideas is rather the object of *epistemé*, science.

So, in describing man's first efforts to speculate about nature, we can say that the mind's eye fell first upon sensible flux and was not held thereby, but turned back to contemplate the world of essences separate from things, the world of eternal archetypes, and thus ended up in what we may call a *metaphysics of the extra-real*. These essences are held to be objects of scientific knowledge,—and of the highest scientific knowledge;—not only are they disengaged but they are existentially separate from reality entire, and placed in a world different from that of things. In short this was a metaphysics of the extra-real conceived by Plato in the image of mathematics. For geometry, too, sets up an extra-real world, and all philosophy that begins in geometry, that blazons the Platonic emblem upon its door, will inevitably be tempted to conceive metaphysics after the type of mathematics, and set the objects of the metaphysician in a separate world.

What, then, do you think will be the result of all this from the point of view of the philosophy of nature? Very simple: there is not and there cannot be a philosophy of nature in a system like Plato's. On the one hand you have *doxa*, opinion, which is concerned with the sensible world and its becoming; on the other hand you have the world of eternal archetypes, the object of metaphysics. On one side you have opinion about the world of becoming and on the other, as science, you have mathematics and metaphysics: no scientific knowledge of nature, no scientific knowledge of the world of movement and time. Wherefore, when the philosopher tries despite everything to give an interpretation of this world and to rise above common opinion, he can proceed only with the help of myths. The use of myths to interpret sensible nature is really indispensable in Plato's philosophy. I think it can be generally said that every attempt to explain natural phenomena by the use of mathematical knowledge alone necessitates the recourse to explanatory myths.

Aristotle

5. Passing on to Aristotle now, what do we find? Aristotle starts by criticising the theory of ideas, by stating that Plato's metaphysics, which we have just briefly dis-

cussed, is false since it is not properly speaking a science but a dialectic. If you will refer to the fourth lesson in Book IV of St. Thomas' commentary on the *Metaphysics*, you will see how St. Thomas, following Aristotle, explains that the word *dialectic* means a knowledge of things by means of logical entities, or *entia rationis*; a knowledge which takes the place of knowledge of things by real causes. Plato's metaphysics is certainly a dialectic in that sense, for it gives us a logical explanation of things and not a real explanation, precisely because the objects it considers are taken under properly logical conditions. That is what we were just saying about the nature of Platonic ideas: they are essences which are separate from things, a state of separation which exists only in the mind.

The metaphysics of Aristotle, instead of being a metaphysics of the extra-real like Plato's, may on the contrary be called a *metaphysics of the intra-real*. Its object is not the world of separate ideas, of archetypes separate from things; it has a wholly different object: being itself *secundum quod est ens*, being taken precisely as being.[3] The object of metaphysics is therefore what the mind perceives to be most inward and fundamental *in* things and not outside of them. This inmost core of things is disengaged for itself, completely disengaged from the sensible, completely disengaged from matter, and this implies that the object thus considered can exist in subjects which are not subject to time and change. The object of thought which the metaphysician calls 'act,' for example, or 'one and many,' etc., can be found realized in non-material subjects as well as in material subjects. This also means to say that, in mutable things themselves, being is not to be considered as mutable, as changing, but precisely as being and, if I may so speak, under its own colors, under its own flag.

The reason why this is so of metaphysics is that the intelligible, instead of being transcendent to things as Plato thought, is immanent to them: it is one of the constitutive elements of reality itself, of the reality which is subject to sensible becoming. That is why, in Aristotle, *ideas* become *forms*. This substitution of the word 'form' for *idea* is of capital importance: we must always keep well in mind that the signification and connotation of the word *form* in Aristotle are totally different from those of the word *idea* in Plato. We sometimes have a tendency to Platonize Aristotelian forms. Although it is true, I think, that Aristotle cannot be understood without Plato as an 'antecedent condition,' yet the more deeply one penetrates into his philosophy, the more clearly does it appear to be fundamentally anti-Platonic: precisely because the intelligible element was completely *de-logicized* by Aristotle. He rid it of the characteristics of an *ens rationis*, an ideal or logical entity, which belonged to it in Plato; he freed it entirely of these characteristics. For his whole philosophy tends toward real existence, whereas Plato's tends toward ideal essence. Instead of being a universal subsisting in an ideal unity, the intelligible element is a spiritual or quasi-spiritual concrete singular, since the form is, in general, something like an adumbration or foreshadow of that which in the living organism is the soul, and in man is the spirit. The intelligible element is a concrete singular which we grasp by means of a universal idea that is in our mind, but insofar as it exists independently of our mind the form is concrete and singular. It is one of the elements of sensible reality itself.

As a result the mind's eye, before attaining in natural things to being as being and its pure metaphysical intelligibility, must and can seize in them an intelligibility which is clothed in the sensible.

For obviously the radical change in the conception of metaphysics which I just mentioned, entails a corresponding change in the conception of the knowledge of nature: henceforward knowledge of a perfect type, solid, scientific knowledge of sensible nature, of change, motion, becoming, is possible. This was Aristotle's great discovery. For us these things have become quite banal, but imagine the unexpectedness and splendor which the flashing forth of these discoveries held for the human mind at the time they were made! Science, scientific knowledge of sensible nature is possible, not indeed insofar as sensible nature is sense-perceived, but insofar as intelligible elements and laws are vested in sense-perceived being. These elements are the natures, laws, intelligible connections and necessities which we must discover and which we can discover under the flux of contingent modifications.

If we may say so, it took great intellectual courage for Aristotle to conquer the mind's temptation to discouragement when faced by the deceptive spectacle of the flight of becoming.

And thus, at a degree much less deep in things than that of being as being (the object of metaphysics), there are disclosed to us ontological diversities and a multiplicity of specific laws in the sensible and changing world. These ontological diversities, this multiplicity of specific laws constitute the object, not of metaphysics, but of what Aristotle calls *physics*, and of what we shall call the *philosophy of nature*.

The point to be remembered here is that Aristotle founded the philosophy of nature in the sense of scientific knowledge properly so called, a science of sensible nature,—which would have been a paradox in the eyes of Parmenides, of Heraclitus and of Plato;—a knowledge of sensible nature whose object is sensible or mutable being taken not as singular or as sensible, but precisely as containing intelligible and universal values which account for its own mutability.

Let us note parenthetically this curious phenomenon: the human intellect, like every intellect, has being for its object and, as human, has for its proportional, connatural object the being of sensible things. The first object of common knowledge, the first object that common prephilosophical knowledge brings out, is what Cajetan calls *ens concretum quidditati sensibili*, being clothed in sensible nature. And yet so ardently does our intellect seek being itself that when, after the stammerings of the first seekers, it set about reflecting in a formally philosophical fashion (for example in the time of Socrates and immediately after him), this intellect which is connaturally ordered to the being of sensible things, discovered metaphysical knowing before it discovered genuine knowing of the sensible and before acquiring a philosophical knowledge of sensible and moving nature. And to begin with it even doubted the possibility of such knowledge and succeeded only with great difficulty in bringing it out. Now that is an extremely suggestive fact. No doubt, the first Greek nature-seekers had prepared the way for Aristotle; that is why he refers to them constantly in his works. But they mixed everything up; they mixed metaphysics and physics and had but a very confused idea of the properly philosophical problems which come up with respect to the knowledge of nature. These problems began to be asked only with Parmenides, Heraclitus and afterwards with Plato. And in a great civilization like that of India, we do not find a philosophy of nature; we find a very rich metaphysics, but no philosophy of nature, or hardly any.

The Orders of Abstractive Visualization

6. How are things organized doctrinally from Aristotle's point of view, from the point of view we have just been talking about and according to which we must distinguish a science of being as being, which is Metaphysics, from a science of sensible and mutable being which is Physics? Here we shall have to refer to that classical doctrine into which we should seek constantly

to penetrate more deeply, for it is truly essential: the doctrine of the three degrees of abstraction or, let us say, the three degrees of abstractive visualization which characterize the three generic types of knowledge.

According to Aristotle and the scholastics we must distinguish between three degrees of abstraction which correspond to the degrees of immateriality or immaterialization of the object. These are the three degrees which permit us to classify the generic types of knowledge. We are told in this classic formula that, in the first degree, that of physics (physics in the general, very universal sense which the word had for Aristotle and which includes, as we shall see, the philosophy of nature as well as the sciences of nature) the mind abstracts from *singular* or *individual* matter only, and the object which the mind presents to itself can neither exist without sensible matter nor be conceived without it; its notion includes material-sensible constituents. This object is being as subject to change: wherefore Aristotle said: "not to know motion is not to know nature." [4]

At the second degree of abstraction we have mathematical knowledge. Here the mind abstracts from *sensible* matter, that is from matter as possessing active qualities perceivable by the senses. The object which the mind presents to itself at this degree is abstract quantity which cannot exist without matter but can be conceived without sensible matter; its notion does not include sensible matter.

Finally at the third degree, we come to metaphysical

knowledge. Here the mind abstracts from *all matter*, from what the ancients called sensible matter (proper to the first order of visualization) as well as from intelligible matter, that is extension, quantity itself which is proper to the second order of abstractive visualization. The object of this metaphysical knowledge is being as being, which can not only be conceived but can exist without matter.

Now I would like to read with you St. Thomas' principal text on this question, which is found in question 5, art. 1, in his *Commentary on the Trinity of Boethius.*[5]

In this text St. Thomas tells us that some of the objects of the speculative sciences are dependent upon matter *secundum esse,* as to their existence, because these objects cannot exist outside of the mind except in matter. But a subdivision of these objects is necessary: for some of them depend on matter *secundum esse et intellectum,* in respect to both their existence and their notion, to exist and to be defined. These things include sensible matter in their definition; they cannot be understood without sensible matter. Thus in the definition of man it is necessary to include flesh and bones. And with objects of this kind *physica* or natural philosophy is concerned. But there are other objects which depend on matter *as to their existence but not as to their notion,* because sensible matter is not included in their definition. Such is the case with line and number; and these are the objects treated by mathematics.

Finally, there are some objects of speculation *which do not depend upon matter for their esse*, their existence, because they can exist without matter. Either these objects of thought, though really existing, are never found realized in matter, as God and pure spirits, or they are sometimes realized in matter and sometimes not. This is the case for the objects of thought: substance, quality, act, one and many, etc. And these two kinds of objects are the objects which are dealt with by metaphysics and theology (theology being taken here in the sense of first philosophy, natural theology).

7. To complete this doctrine we must say a few words about a distinction which is very important in scholastic thought and which is not usually given enough stress: the distinction between what is called *abstractio totalis*, abstraction of the whole with respect to the parts (we shall call this extensive abstraction) and *abstractio formalis*, abstraction of the form or formal type from matter (which we shall call intensive or typological abstraction).

What the Thomists call "abstractio totalis" is the extraction of the universal whole considered as such, abstraction by which we draw the object of thought "man," for example, from Peter, Paul and John; the object of thought "animal" from man and so on, proceeding in this way to more and more general and larger universals. The point of view here is that of greater or lesser generality and this extensive visualization, this abstraction of the universal whole, is common to all knowledge, to pre-scientific knowledge as well as to the scientific knowledge by which it is presupposed.

The other type of abstraction on the contrary, abstraction of the formal type, typological visualization, consists in the extraction of the intelligible type by which we separate what belongs to the essence or formal *ratio* of an object of knowledge from the contingent and material data.

This is Cajetan's doctrine; we find his exposition of it in the *Prooemium* of his Commentary on the *De Ente et Essentia*, question 1. But in order to avoid a possible misunderstanding or a verbal difficulty we should note right away that there is a difference of vocabulary here, between St. Thomas and Cajetan. In his *Commentary on the Trinity of Boethius*, St. Thomas distinguishes mathematical abstraction from physical abstraction in the following terms: he says that mathematical abstraction corresponds (to disassociate it) to the union of form and matter, or more precisely to the union of accidental form and its subject. This is the way the abstractive eye of the mathematician abstracts the accidental form we call quantity from the material subjects in which it is found: *abstractio formae a materia sensibili*: he leaves aside sensible matter so that he may consider only the accidental form *quantity* separately from corporeal substance.

Physical abstraction, on the other hand, corresponds (to disassociate it) to the union of the whole and the part; this is the abstraction of the universal from the particular, *abstractio universalis a particulari*; abstraction in which the abstractive eye of the physicist considers a certain nature by itself, according to its essential *ratio*, separately from all the parts which are not constitutive of the species but are accidental parts with respect to it.[6] That is the way St. Thomas states the case. He calls *abstractio formalis* the second order of abstraction (mathematical abstraction) and he calls *abstractio totalis* the first order of abstraction. Whereas for Cajetan, the first order of abstraction, physical abstraction leading to a *scientific knowledge* of nature, is itself an *abstractio formalis*, as is abstraction of the second and third order. For him *abstractio totalis* is simply a general condition of human knowledge, prerequisite to science.

As a matter of fact, what we have here is a simple difference of vocabulary, not a difference of doctrine. For St. Thomas the first order of abstraction considers the nature of a thing separately according to its essential *ratio*, disengaging it from the parts which are only accidental in respect to the specific essence. Now such a process constitutes an *"abstractio formalis"* in Cajetan's sense of the term, the only difference being that in the first order of abstraction the form considered separately is the very nature itself, the specific essence; whereas in mathematical abstraction the form is an accidental one separated from the subject: not the human nature of Peter, Paul or John, but the accidental form *quantity*, separated from corporeal substance. It is this difference between the first and second orders of abstraction that

St. Thomas was stressing in the text of his *Commentary on the Trinity of Boethius* which we have just quoted; but both cases are instances of "abstractio formalis" in Cajetan's sense of the term.

I would like to insist on this point: when I say "circle" "straight line," "the number two," evidently I am abstracting a form from a subject or a matter and I am separating this form from the accidents which may belong to it in such and such of its material subjects. In reality, a circle is colored, made of wood or iron, etc. These are accidents with respect to the form *circle*, accidents which I separate from that form in order to consider it in itself. Likewise duality belongs in reality to two yards of cloth or to two soldiers in a regiment, accidental conditions with respect to the intelligible type presented by the concept *two*; I separate off this intelligible type, leaving aside the material accidents to which it is united in concrete materiality. Mathematical abstraction in which we separate the accidental form, quantity, from the subject in which it inheres, offers us a perfectly clear example of *abstractio formalis*.

But if, on the contrary, I place myself at the first degree of visualization, the physical degree, and say "man," "rational animal," these words indicate both *abstractio totalis* which is pre-scientific, and *abstractio formalis*; yet there is an essential difference in the manner in which I think in each instance. (This point demands close attention for it can be rather confusing.) I use the same words but the act of thinking I perform is different in the one case and the other. In the first case, the case of *abstractio totalis* or extensive visualization, I simply abstract the universal whole from the parts. I could just as well say, instead of rational animal, "featherless biped" or "monkey-metaphysician." If I disengage the essence exactly so much the better for me, but it is not precisely the essence as such that I would attain to in this sort of abstraction; I am simply trying to reunite the common traits, to set up a simple notional framework common to such and such individuals, Peter, Paul or John. In the second case on the contrary, (*abstractio formalis* or typological visualization), when I say "rational animal" this same word corresponds to a wholly different act of thought. Here I am trying expressly to attain to the nature, the essence, the type of being, the locus of intelligible necessities; in brief to the object of *science* discernible in these individuals, Peter, Paul or John. So you see, although I have been using the same word "man" or "rational animal" in both cases, I have been dealing with two very distinct acts of thought.

To illustrate this difference, to clarify it a bit, let us say that in the first case, the case of extensive visualization, the word "Catholic" for example, evokes an average and purely empirical notion uniting all the traits common to a certain number of individuals of this denomination; for example the notion established by *taking an average* of the subscribers to *La Croix, Le Pèlerin, La Vie Catholique,* even *L'Echo de Paris.* But in the other case (typological visualization) this word has another connotation: it designates the visible members of the Mystical Body of Christ, called by baptism to sanctity. In one case the mind is expressly trained upon the intelligible type; in the other it simply aims at an average, a common mark or trait.

8. This doctrine has really been clarified only by Cajetan in the *Prooemium* of the *De Ente et Essentia*. "In order better to understand these things," writes Cajetan, "we must note that, just as there is a double composition in things, *i.e.* of form with matter and of the whole with the parts, so is there a double abstraction performed by the intellect: one by which the formal is abstracted from the material, *quo formale abstrahitur a materiali*, and the other by which the universal whole is abstracted from the subjective parts. According to the first kind of abstraction quantity, for example, is abstracted from sensible matter; according to the second, the generic universal "animal" is abstracted from bull or lion. The first sort of abstraction we call *formalis*, the second we shall call *totalis*."

Then Cajetan explains the differences between these two types of abstraction, noting that typological visualization proceeds actuality-wise, in the direction of distinction and intelligibility, whereas extensive visualization proceeds potentiality-wise, towards the least intelligibility. This is why,—(another difference),—in typological visualization the more an object is abstract the more it is *known in itself (natura)*, whereas in extensive visualization the more an object is abstract, the more it is *known to us.* "The basis for this difference is that, in abstraction of the formal type, potential material elements, etc., are put aside, whereas in abstraction of the universal whole, on the contrary, specific actualities are put aside. And that is why, in this latter kind of abstraction, the more an object of thought is abstract, the more it is potential, since it is in potency that the genus contains its inferiors." [7]

Furthermore, Cajetan adds,—and this is of great import for us—: "The speculative sciences are distinguished one from the other according to the different modes of *abstractio formalis,* (typological visualization,—and not according to *abstractio totalis* which is simply pre-supposed by science). But *abstractio totalis,*—extensive visualization—, is a condition common to all the sciences whatever they may be." And he points out this consequence which is very important for an authentic understanding of metaphysics: "That is the reason why the objects of metaphysics as such are not compared to the objects of the physicist as a universal whole to subjective parts," to more particular objects of thought; the difference between them is not a simple difference of extension, as the modern philosophers are always insisting.

The being which is the object of the metaphysician does not differ from the being which is the object of the physicist merely because it is more common; in fact the greater the tendency toward the common as such, the greater the tendency toward the potential, the indeterminate, the merely more common being which could not be an object of science. The objects of the metaphysician are compared to those of the physicist *"ut formalia ad materialia,"* [8] as the formal to the material. It is a purer form abstracted from matter, an intelligible reality of a superior type and *surordinate* to the others. "Although intelligibles of the metaphysical degree are more universal than others and can be compared to them as to parts contained in their extension, nevertheless, insofar as they are metaphysically considered, they are not (wider, more common) universals with respect to the objects of the physicist (the philosopher of nature), they are (regulative) forms." [9] Which is why metaphysics has a regulative function with respect to the natural sciences, just as mathematics also has a regulative function in relation to them, because metaphysics' and mathematics' object is not merely more universal, vaster, wider, it is a form, a pure type abstracted from material conditions.

This doctrine is of capital importance: it shows how the criticism levelled at metaphysical abstraction by so many modern philosophers, be they named Brunschvicg or Blondel, is based on ignorance of the question. Brunschvicg, for example, accuses the "pre-Cartesians" of explaining a thing by its idea or general notion, by the logical framework wherein the mind has placed it and which the mind more or less hypostasizes or ascribes reality to. Now that, in the eyes of a scholastic, would be a purely logical explanation and worthless from the scientific point of view; it would be a dialectical explanation, to use the Aristotelian word mentioned above, an illusion or a dream of science, but in no way science itself. Yet this is the only and exclusive type of abstraction which the philosophers I am talking about think of; these authors conceive only of *abstractio totalis*, extensive abstraction, and therefore they attack a fancied abstraction which has supposedly scientific pretensions while remaining in actual fact *abstractio totalis* and never reaching the level of *abstractio formalis*, of typological abstraction. Thus they are fighting chimaeras and one might think them to be as ignorant of the true conditions of science as the supposed adversaries they criticize. For Thomists, as Cajetan has just shown us, science begins with *abstractio formalis*: before that there is no science; there can only be common or vulgar knowledge but no science, nor any perception of intelligible necessities.

Since he conceives of no other *philosophical explanation* than a logical or dialectical explanation by a being of reason (*ens rationis*) or ideal entity, it is natural that a nominalist philosopher like M. Brunschvicg should renounce philosophical explanation thus conceived and accept only mathematical or physical explanations which he terms the sole "rational" ones. But Thomists can certainly return the compliment to these philosophers. You know that according to certain idealist theorists concerned with the diverse ages of the intellect, Thomism corresponds to the mental development of a seven to nine year old child. Well, their own conception of the life of reason applies exactly enough to the mental development of a child who has grown up without becoming an adult and who nevertheless pretends to science, to the activity of a grown man without ever having attained,—except in mathematics,—to the proper means of science, that is to the typological visualization we have just been talking about.

9. But enough of this parenthesis; let us return to our subject and to the degrees of abstraction. What I would like to impress upon you is that there is not a simple difference of degree, nor a simple difference of generality between these orders of abstraction; they are not in the same generic line one above the other. When, in order to characterize the proper object of the sciences which are classed within these degrees, we say that at the first degree we consider sensible or mobile being, at the second degree quantified being, at the third degree being as being, we must note well that the word "being" (*ens*) has, in each of these three cases, an *analogical* meaning. The division in question is an *analogical* one: the word and the concept *being* are not used in the same manner in these three cases. We must insist upon this point because there is an unfortunate tendency among philosophers to misunderstand analogy, to treat an analogical concept like a generic concept; to reduce the differences between the analogates of an analogical concept to simple differences of degree in the same line, within the same generic concept.

What we must bear well in mind is that the three kinds of abstraction, physical, mathematical and metaphysical (which are degrees of typological visualization)

answer to essentially different types of intellective operation. There is an essential heterogeneity between these
different degrees. That is why St. Thomas teaches, in
his *Commentary on the Trinity of Boethius*,[10] that the
aim or term of knowledge (which pertains to judgement, for it is in the judgement that cognition is per-
fected) *is not always of the same kind* in the different
types of speculative knowledge. Physical knowledge terminates in the sensible; mathematical knowledge terminates in the imaginable; metaphysical knowledge in the
pure intelligible. And we have here a precious text which
should be written in letters of gold over the portals of
every university: "In things divine (metaphysical) we
should not be brought, as to the term in which we verify
our judgements, either to the sense or to the imagination.
In the case of mathematical objects we must verify our
judgements in the imagination, not in the senses [granted
of course that this verification must itself be understood
in an analogical, often indirect manner, as in the non-
Euclidean geometries]; but in the case of the object of
physics, knowledge terminates in the senses themselves;
it is in the sense that the judgement is verified." And
St. Thomas adds: "Wherefore it is an intellectual sin to
want to proceed in the same manner in the three divisions of speculative knowledge." [11] That was the sin of
Descartes, who wanted to reduce all the speculative sciences to one same degree, one same method, one same
type of intelligibility.

There is another interesting text on this subject in the

VIth book of the *Metaphysics*, First lectio X, by St.
Thomas. If you consult this text you will see that the
same doctrine is set forth there as in the *Commentary on
the Trinity of Boethius*: "Et universaliter omnis scientia
intellectualis qualitercumque participet intellectum: sive
sit solum circa intelligibilia, sicut scientia divina; sive sit
circa ea quae sunt aliquo modo imaginabilia, vel *sensibilia in particulari, in universali, autem intelligibilia, et
etiam sensibilia prout de his est scientia*, sicut in mathematica et in naturali: sive etiam ex universalibus principiis ad particularia procedant, in quibus est operatio, sicut
in scientiis practicis: semper oportet quod talis scientia
sit circa causas et principia." [12]

Note this phrase: *et etiam sensibilia prout de his est
scientia*. The object of physics falls under the senses precisely as an object of science; sense perception plays an
essential role in the knowledge proper to the first degree
of abstraction (in a manner which we have yet to explain
and which we shall try to make clear later on).

10. Now let us further discuss this typical heterogeneity and say a few more words about the three orders
of abstraction.

At the first degree of abstractive visualization our
mind deals with intelligibility involved or clothed in the
sensible itself. A certain intelligible being is doubtless
present, disengaged from the singular and contingent
moment of sensorial perception, but this being is not
disengaged for itself and in all its amplitude; it is particularized according to the diverse natures of the sensi-

ble world. And doubtless there are many specific degrees within this same generic universe of intelligibility that we, with Aristotle, are calling here the universe of the "physica."

But as long as we stay in this universe of the first order of abstraction, knowledge, however illumined it may be by the very intelligibility of being, however decanted it may be, remains held within the limits of sensible existence, of sensible mutability and its causes. For here intelligibility itself implies a reference to the sensible; the definitions we give always bring the mind back,—in an essentially varied fashion as we shall see later on,—to certain sensible *data* which it receives from experience and beyond which it cannot go. The mind both depends on and finds its limit in this sensible data; it has to accept it; this is a condition of humility which the philosopher of nature as well as the scientist must accept. We are dealing here with *intelligible being involved in sensible existence*; let us call this sphere of intelligibility, this universe of knowledge, the realm of the *sensible real*. One might say that here the mind plunges into the ocean of the sensible in order to seize in this ocean of the sensible itself the intelligible structures which exist therein. Here the object of knowledge is being as mobile, being under the typical determination of mutability, being as imbued with mutability.

At the second degree of abstractive visualization we have to do with a wholly other universe of intelligibility. This second degree is not on the same line as the first, for here being is disengaged from the experimental sensible as attained by the external senses. But as St. Thomas stressed in the text we have just read, it remains clothed in the imaginable either directly or indirectly. At this degree intelligibility no longer implies an intrinsic reference to the sensible, but to the imaginable. Here instead of seeking the intelligibility of the sensible real, the contact of the mind with the sensible is for the mind the occasion of separating certain intelligibles which it extracts from the real and which it considers outside of the whole order of reality, outside of the whole order of possible existence. To use a metaphor, we might say that the mind, here, is like a sea-bird which gets hold of a fish and then soars into the air to eat it; it does not penetrate into the ocean of the sensible but it gets hold of an intelligible to devour it in another milieu.

In this case we are dealing with an intelligible sphere which is not that of the sensible real and which we may call the sphere of the *mathematical preter-real*. And the reason why the mind can lay hold of these intelligibles in this way and with them set up a world apart, a separate universe of knowledge, is that sensible matter is not implied in the concept of these intelligibles. Here the object of knowledge is not being as imbued with mutability, as it is in the first order of visualization, nor being as being, as it is in the third order, but it is a certain particular being constituted by essential forms and the relations of order and measure peculiar to quantity; in a word it is being as quantity, (as that particular being which is called quantity). Here the shadow of matter still darkens intelligibility in a certain manner for, in the very first definitions which are at the origin of science, the object in question is not defined except by implying imaginable elements; not by implying sensible elements, sensible matter, as in the case of the first order of abstraction, but imaginable elements in which the mind finds its limit and beyond which it cannot go.

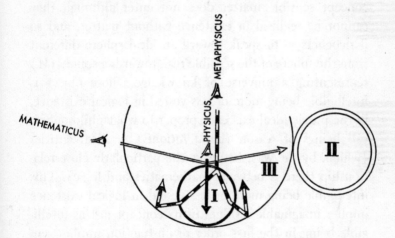

We can present all this in an illustrative diagram. Let a sphere (I) represent the first order of abstraction. The intellectual eye of the "physicus" falls on the sensible surface and penetrates into the sphere where, supposing this sphere to be heterogeneous, it finds different strata of intelligibility. Starting from sensible phenomena it plunges deeper and deeper into the ontological depths within this universe of knowledge. This is the sphere of the sensible real, the sphere of intelligible being more or less vested in the sensible; 'more or less' for this sphere is not homogeneous; there are specific differentiations within this universe of knowledge.

Now in the second order of abstraction, the mathematical order, let us say that at the moment in which the mind's eye falls upon the sphere of the sensible real, it discerns therein intelligibles of another kind into whose concept sensible matter does not enter although they cannot be realized in existence without matter, and so it ricochets so to speak toward an ideal sphere different from the sphere of the sensible real, towards a sphere (II) representing a universe of knowledge whose object is intelligible being more or less vested in logical existence, in the purely ideal existence proper to what philosophers call beings of reason (*entia rationis*) or ideal entities (which, by the way, bespeaks the particularly close relationship there is between mathematics and logic). This intelligible being more or less vested in logical existence implies imaginable residues in its concept just as intelligible being in the first order of abstraction implies sensible residues in its concept. So the word "*ens*" has a wholly different bearing in one case and in the other.

To come finally to the third degree of abstraction, suppose that the philosopher's intellectual glance stays *in the real* in order to fathom it and that, being refracted in the sphere of the sensible real, it passes *beyond* this sphere and discovers a third, much vaster universe (III) that may be called the sphere of the pure intelligible, or again the realm of the *trans-sensible real* (sphere of the metaphysical trans-sensible which itself opens on to the analogical knowledge of trans-intelligible objects). It is as if, by dint of diving ever deeper into the ocean, one finally succeeded in finding at the bottom of the sea a magical mirror reflecting the sky. The glance is thus reflected above towards purely intelligible objects and this is the glance proper to the metaphysician, to metaphysical visualization. What we are dealing with here is real being (just the opposite of what happens in mathematical abstraction), real being disengaged from sensible existence; no longer vested in it as in the first order of abstraction but disengaged from sensible existence, grasped for its own sake in an original intuition. Being as being, being "sub ratione entitatis" or as merely connoting *esse*, has been freed of any sensible or imaginable gangue so that it may be considered in its pure intelligible type. The word "ens" in this third case has only an analogical community of meaning with the word "ens" as used in the two other cases, "ens quantum" or "ens sensibile seu mobile."

Metaphysics, the Philosophy of Nature and the Natural Sciences

11. From these explanations of the three orders of abstraction we may conclude,—and this is an essential point,—to a capital truth which has been brought out since the time of Aristotle, a truth to which we must

always adhere: *there is an essential distinction between the philosophy of nature (or "physics") and metaphysics.* The sphere of the first order of abstractive visualization, which Aristotle called φυσική and St. Thomas "philosophia naturalis," includes in its extension the philosophy of nature as well as the experimental sciences of nature. The philosophy of nature is essentially distinct from metaphysics; theirs are two different universes of intelligibility. The intellectual glance of the philosopher is different in the one case and the other: the primary intuitions we are dealing with here are basically different.

In the case of the philosopher of nature we have an intuition of being particularized in sensible natures, of being imbued with mutability. In the case of metaphysical intuition on the contrary we have being taken in all its fullness, in its very intelligibility as being; whereas for both the philosopher of nature and the scientist, being is taken in the inferior and darkened intelligibility which befalls it as a result of its vestiture in the sensible. There are some young teachers of scholastic philosophy who think that the philosophy of nature does not exist as a discipline essentially distinct from metaphysics; they would like to absorb the philosophy of nature into metaphysics. In this they sin against both St. Thomas and Aristotle; they are unwittingly followers of Wolff.

This distinction of natural philosophy and metaphysics must be considered as absolutely fundamental because it relates to the first intuitions of being. We can seize being intuitively either *as being* in all its intelligible

purity and universality, or as *involved in the sensible*. To
Aristotle goes the credit for having clearly brought out
this essential distinction from the beginning, a distinction which is linked to the very birth of the philosophy
of nature. We saw above that at first metaphysics tried
to set itself up without the philosophy of nature, or to
the latter's exclusion, and that Aristotle's achievement
was to maintain and definitively to constitute metaphysics the while he made room beside it for knowledge
of the sensible, for a science of nature itself. In that
science being is known precisely as sensible, as mutable;
which fact essentially distinguishes this universe of knowledge from the metaphysical universe.

12. However the ancients, Aristotle himself and the
early scholastics paid for this capital truth by a serious
error of intellectual precipitation. We cannot say that
the ancients were incurious about the details of phenomena; they were just as interested as the moderns are,
but they failed to perceive that this detail of phenomena
needs a science of its own, its specific science, specifically
(I do not say generically) distinct from the philosophy
of nature.

In the optimistic view of the ancients, who were
prone to arrive quickly at what were often-times very
hypothetical or fallacious explanations as concerned the
detail of phenomena, philosophy and the experimental
sciences were one and the same science and all the sciences concerned with the material world were subdivi-
sions of *one unique specific science* which was called "phi-

losophia naturalis" and to which it belonged at once to analyze the nature of corporeal substance and to explain rainbows. For the ancients, the philosophy of nature *absorbed all the natural sciences.* For them the detail of phenomena was not the object of a specifically distinct scientific explanation.

No doubt they distinguished different degrees within the philosophy of nature: they distinguished therein the science of *quia est* (ὅτι) and the science of *propter quid* (διότι), that is to say the science which is concerned with simple verification of fact and the one which assigns reasons for facts and is deductive in type. But these were divisions of one *same,* more or less perfect, speculative science. For them there was just one single specific science of nature with different degrees. As far as the science of phenomena is concerned, the ancients lacked a certain conceptual equipment, a certain conceptual technique. Not only did they lack this or that laboratory equipment but they also lacked the appropriate conceptual technique. They discovered the intellectual instrument of analysis of natural phenomena only in certain special fields, such as astronomy and optics, (for they by no means ignored it entirely), but they did not conceive of the possibility of a general science of sensible phenomena specifically distinct from the philosophy of nature. Outside of the special fields we have just mentioned, astronomy, optics, harmony, the study of phenomena was limited for them to very general interpretations conducted under the light of philosophy, essentially ordered to an ontological knowledge and analysis of things; interpretations which were less certain, more dependent on simple probability and closer to simple opinion, the nearer they came to the detail of phenomena.

For the ancients the analysis which we shall later call the *ontological* type absorbed all other types.

And that continued to be the case in the Middle Ages and right down to the seventeenth century. In the ancient treatises on natural philosophy, in the *Cursus philosophicus* of our master, John of St. Thomas, you will find long discussions about meteors, explanations of rainbows, snow crystals, etc. Likewise in the first editions of Goudin there was, I believe, a formal refutation of the pneumatic machine. . . .

For the scholastic philosophers, the questions which today are called scientific did not constitute a specifically distinct discipline, but were just a chapter of philosophy. And note well that this was true of Descartes too. Descartes was the source of a distinction which he himself did not make: this distinction was made from his time on, but Descartes himself considered that he was writing a chapter of philosophy when he wrote a book on meteors. On this, see Gilson's essay on "Cartesian Meteors and Scholastic Meteors." [13]

To absorb all the sciences of nature into the philosophy of nature was an error in the speculative realm and we are much indebted on this point to the work of modern times for a historic gain which the Thomistic synthesis must always take into account.

SECTION 2 · THE GALILEO-CARTESIAN
REVOLUTION

The Intermediary Sciences

13. Our first chapter is concerned with Greek and mediaeval philosophy and its difficulties. It is into these difficulties that we are now going to look: they cropped up with the Galileo-Cartesian revolution. At the end of this revolution we shall witness the inverse error to that made by the ancients: the ancients absorbed the sciences into the philosophy of nature; the moderns will end up by absorbing the philosophy of nature into the natural sciences. A new, inexhaustibly fecund discipline will have established its rights. But this discipline, which is not wisdom, will have supplanted wisdom,—the wisdom *secundum quid* of the philosophy of nature and the higher wisdoms.

There, below the metaphysical plane, in the world of the first order of abstraction a hidden drama was played between Physico-mathematical Knowledge and the Philosophical Knowledge of sensible nature; a drama whose consequences were of primary importance for metaphysics itself and for the intellectual regimen of humanity. This drama had two principal acts: in the first, physico-mathematical knowledge was mistaken for a philosophy of nature, for *the* philosophy of nature; in the second it wholly excluded any philosophy of nature.

14. The first act lasted for two centuries, from the time of Galileo and Descartes to that of Newton and

Kant. At the beginning of the XVIIth century a new astronomy, physics and mechanics triumphed over the explanations of the detail of phenomena that were being taught in the name, alas, of the philosophy of Aristotle. Prepared by the researches of the great scholastic scholars of the XIVth and XVth centuries, foreshadowed and as it were foretold by Leonardo da Vinci and certain other Renaissance thinkers, a new epistemological type, a new type of conceptual equipment found its way into thought: a type which consists above all in a mathematical reading of the sensible.

This science, which has been so vastly successful these last three centuries, may be said to consist of a progressive mathematization of the sensible and its success, as you know, has been especially remarkable in Physics. The type to which it belongs was not unknown to the ancients but they had discerned it only in restricted and particular fields such as astronomy, harmony and geometric optics. They had pointed out that these were intermediary sciences, *scientia media* as they so aptly termed them. According to Aristotelian and Thomistic principles, such knowledge must be considered as formally mathematical because its rule of interpretation, its rule of analysis and deduction is mathematical.

But on the other hand, although this knowledge is formally mathematical, it is materially physical because what it assembles and interprets by the help of mathematical intelligibility (particularly, as soon as this tool was invented, by the help of a system of differential equations) is physical reality, physical data. Such sciences may hence be said to be formally mathematical and materially physical. They are, as it were, astride the first and second orders of abstractive visualization; materially they belong to the first order; formally,—and this is what is particularly important in a science: its rule of explanation and interpretation,—formally they belong to the second order of abstraction, to the mathematical order.

One further remark about these sciences. In his Commentary on the second book of Aristotle's *Physics*, St. Thomas draws attention to the fact that, while these sciences are formally mathematical they are nevertheless more physical because, says he, their term,—the terminus in which judgement is completed and verified,—is sensible nature.

Herewith an historical parenthesis. As a matter of fact what I just mentioned was not expressly said by Aristotle; St. Thomas said it, basing himself on a text of Aristotle which, to our greater advantage, he misinterpreted. In Chapter 2, 194a, 7, lib. II of the *Physics*,[18] Aristotle is speaking of mathematical knowledge and he speaks of the branches of mathematics which are more physical than others, more concerned with physical things; these he calls τὰ φυσικώτερα τῶν μαθημάτων, which modern translators have with good reason translated as "the more physical of the branches of mathematics." St. Thomas, on the contrary, in his 3rd lesson on the second book of the *Physics*, understands this expression to mean not the more physical branches of mathematics but sciences that are more physical than mathematical, *magis naturales quam mathematicae.* That enables him to state a very important point of doctrine, namely that the while they remain formally mathematical, these sciences are more physical: *quia harum scientiarum consideratio* TERMINATUR *ad materiam naturalem, licet per principia mathematica procedat.* Their weight as science draws them toward physical existence although their rule of interpretation and deduction is mathematical. From this you can see immediately the kind of internal disparity there will be in this scientific realm: the scientist will be drawn at the same time towards the physical real with its proper mysteries and existentiality, and towards mathematical intellection and explanations. At certain times one or the other tendency will appear to predominate; actually the scientist is on both these planes simultaneously; he is the more on the physical plane as to the term of knowledge, the while he is on the mathematical plane formally, as to the rule of interpretation and explanation.

To come back to the diagram we used above to illustrate the three degrees of abstraction (section 1,n° 10). In it we drew a sphere representing the first order of abstraction, wherein the mind's eye looks more and more deeply into the sensible real. Then there was another vaster realm which is that of metaphysics or the third order of abstraction, in which the mind's eye having penetrated the sensible real is reflected above toward

suprasensible realities. In the second order of abstractive
visualization the intellectual glance, having fallen upon
the sphere of the sensible, detaches from it by mathematical abstraction certain intelligible natures (which
do not imply sensible matter in their definition) and
ricochets toward another sphere which we have called
the sphere of the mathematical preter-real,—because the
mathematician is not interested in the order these entities have toward existence.

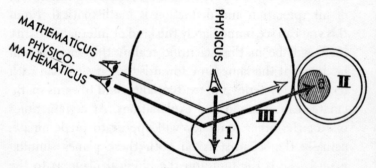

Now to illustrate what we have just been saying about
the *scientiae mediae,* we shall represent the mind's eye
as penetrating to the *interior* of the sphere of the sensible
real; but instead of remaining purely and simply in the
sensible realm, its glance is refracted toward the mathematical sphere: and thus we have in the mathematical
sphere a projection, as it were, of this sensible domain,
we have this mathematization of the sensible which I
have been talking about and which constitutes physico-
mathematical science. This smaller sphere (a) belongs
in the mathematical sphere because the science in ques-

tion is formally mathematical, although materially physical.

A Tragic Misunderstanding

Thus, in order to interpret the whole field of natural phenomena, the new type of knowledge concerned itself with sensible reality, sensible and mutable being as such, from which it started and to which it returned, but it set about viewing it not under its ontological aspect but under its quantitative aspect: it set to deciphering it rationally by means of the science of the continuous and of number. Clearly the result of this approach could not be a Philosophy of nature but very precisely speaking a Mathematics of nature.

A true estimate of what essentially constitutes this physico-mathematical knowledge shows us that it was very foolish for the scholastics of the decadence to oppose it as if it were a philosophy of nature contrary to their own. But it was also very foolish of the moderns to ask that this mathematical reading of sensible phenomena speak the last word about the physical real and to consider it as a philosophy of nature opposed to that of Aristotle and the scholastics. Hence this great epistemological tragedy was based on a misunderstanding. This sort of thing is always happening, this sort of historical knot contingently produced between essentially different energies (here between different epistemological types) and resulting in an intellectual dispute about a badly stated problem. At that time it was almost inevitable. Now, after long historical reflection it is easy to size up the situation; but at the time this physico-mathematical knowledge was introduced into the sphere of the sciences it was difficult not to mistake it for a philosophy of nature. So the problem was posed in the same way for both the scholastics and their adversaries,—and in an erroneous way. Both of them thought themselves faced with a choice between the old philosophy of nature and the new; but actually there was on the one hand a philosophy of nature and on the other a discipline which *cannot be* a philosophy of nature: two sciences which do not fish in the same waters and are therefore perfectly compatible.

Now a mathematical interpretation or reading of the sensible is, of course, possible only with the help of the basic mathematical notions, with the help of geometric entities and number (and necessarily of movement too; although movement is not of itself a mathematical entity it is an indispensable intrusion of physics into mathematics). So, obviously, from the moment physico-mathematical knowledge of nature was mistaken for a philosophy of nature and was asked to give an ontological explanation of the sensible real, the human mind was bound inevitably to tend toward a mechanistic philosophy and to endeavor to explain everything,—in the philosophical sense of the word explain,—in terms of extension and movement. It was bound inevitably to endeavor to make ontological reality intelligible in terms of extension and movement.

You can see how Descartes' rigorously mechanistic philosophy of nature was,—and this indeed is what condemns it as a philosophy,—a marvelously servile adaptation of philosophy to the dynamic state of the sciences and of scientific research during his time. He transferred into the philosophic order the very outlook that science needed from its methodological point of view and in the physico-mathematical order. As a result, since science aims at giving a mathematical interpretation of sensible nature, it will be thought that science,—confused with the philosophy of nature—, must explain the whole of ontological reality by extension and movement. Well, if science cannot do so right away it will be able to later on; but in any event this necessity is inscribed in its nature. It will be thought that every phase of knowledge in which things are not explained in this fashion must be considered as provisional; that in this provisional state, philosophical thought is not yet fully itself, and that, if it is itself, it is so to the measure in which it approaches an integrally mechanistic explanation.

To return to our diagram: the error we are talking about consists in thinking that the sphere (a) which represents physico-mathematical knowledge and which we have placed within the mathematical sphere, represents the sphere of the philosophy of nature itself. That sets physico-mathematical knowledge up as philosophy of nature. Now, as regards our way of knowing, the philosophy of nature inevitably holds the place of basic wisdom, the most imperfect wisdom which we begin with, in the organic structure of human wisdom. Consequently physico-mathematical knowledge mistaken for philosophy of nature becomes the primary center of organization for all philosophy, and it is around this so-called philosophy of nature, confused with physico-mathematical science, that metaphysics will be constructed. From this we can see how metaphysics has been led astray since the beginning of the XVIIth century; for all the great systems of classical metaphysics since Descartes have taken as ground-floor key to the system of philosophical knowledge a so-called philosophy of nature which was the mechanistic hypostasis of the physico-mathematical method.

II

The Positivistic Conception of Science, and Its Difficulties

SECTION 1 · THE POSITIVISTIC CONCEPTION OF SCIENCE

The Genesis of the Positivistic Conception

1. We have been speaking of a first phase in the vicissitudes of the Philosophy of Nature during these modern times. There was to be a second phase which began in the 19th century and continues to our day.

In the first place, it was evident from the beginning and after several vain attempts at integral materialism it became even clearer, that the things of the soul and even those of organic life (Descartes notwithstanding), are irreducible to mechanism.—Descartes himself was well aware of this as concerns the things of the soul; that is why he paralleled the absolute mechanism of his corporeal world by an absolute spiritualism for the world of thought.—Despite many efforts this dualism was never overcome: which is not a good sign for knowledge that

45

pretends to be philosophy. So there was already a vast intelligible sphere that escaped mechanistic philosophy, the explanation of everything by motion and extension.

In the second place, and this took a long time, science gradually became aware of itself and its processes. This law of self-awareness is a general law for all spiritual activities but since man is not a pure spirit,—being a rational animal he most often thinks "on the level of sensation,"—this law normally takes a long time to operate. It is not surprising that physico-mathematical science took about three centuries to become aware of itself. It became aware of itself and of its processes bit by bit and thereby freed itself of the philosophical and pseudo-philosophical gangue that mechanism had surrounded it with; as it became self-aware it perceived increasingly that it was not a philosophy.

Finally, in the third place, we must keep in mind the influence of the Kantian critique (and note that from this point of view Kant's work was most meritorious, if it be limited by abstraction to the epistemological considerations we are here discussing). The Kantian critique showed that the science of phenomena, what is called "science" in modern parlance, is not equipped to discover the thing in itself, the cause in its ontological reality. This inability on the part of experimental scientific equipment to pass over to metaphysical or more generally, ontological, philosophical knowledge may be considered as one of the basic intuitions of Kant's thought. Kant saw this inability very clearly; his error was to want to generalize this view, to draw from it a whole system concerning the nature of knowledge in itself.

2. Under the influence of the three factors just pointed out, physico-mathematical knowledge of nature, which in the XVIIth century had been mistaken for ontology and for the philosophy of nature, was gradually brought back to the interpretation of phenomena. Its essential goal was, —and this is absolutely correct,—to construct a texture of mathematical relationships, deductive in form, between observable and measurable phenomena. So in the XIXth century experimental science became what it had always unwittingly been: a science of phenomena as such. On this point we are indebted to Kant for having brought into philosophical usage the word *phenomena*, not as he uses it in his own theory of knowledge but insofar as it very exactly qualifies the epistemological realm we are talking about, the science of phenomena as such essentially distinct from ontology properly so-called, from the ontological and philosophical knowledge of nature. This is a very important truth and well worth keeping in mind.

At the same time, from the beginning of the XIXth century on, this liberation from philosophical preoccupations and pretensions, key-noted by physics under the impulsion of mathematics, was extended bit by bit to the whole realm which we shall soon call *empiriological*, even to the sciences of phenomena which do not constitute a physico-mathematics, to the sciences of phenomena which do not yet admit, or which can never admit, of mathematical interpretations.

So on our diagram we may draw a sphere (b) to represent this realm interior to the knowledge of the sensible real and constituting a non-mathematicized sphere of the knowledge of phenomena as phenomena, distinct from the philosophical interpretation of nature. The science of phenomena as phenomena thus comprises two differ-

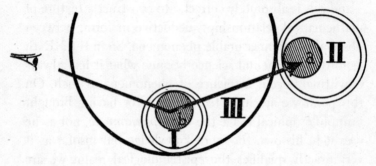

ent types that are actually often inter-mixed: first, physico-mathematical knowledge, mathematical interpretation of the sensible and second, knowledge of phenomena linked phenomenon to phenomenon without mathematical interpretation or with a curtailed mathematical interpretation which is incapable of giving to knowledge its deductive form.

Knowledge of phenomena which are linked phenomenon to phenomenon, yes! But not based philosophically upon an idealist substructure, as Kant thought, but based rather on a (implicit or more or less subconscious) *realist* substructure, as Meyerson was to point out; realist in this sense that this intelligible linkage of phenomenon to phenomenon presupposes the reality of things and is established by getting back to this reality by means of entities constructed by reason, or of explanations and causes constructed in the manner of *entia rationis*.

The Advent of Empiriological Thought and the Concept of Science

3. We have said that the experimental sciences definitely became sciences of phenomena and that this is an important truth, one well worth keeping in mind. Thus was set up for itself and according to its own laws a universe of science which is in no way, not even *secundum quid*, a wisdom. In itself such a differentiation bespeaks considerable progress. But this progress had its other side and had to be paid for: at that moment the sciences in question began to pretend to absorb all the knowledge of nature, to claim for themselves alone the knowledge of sensible nature, with the result that only one knowledge of nature, one sole science of the physical world was accessible to man: this science of sensible phenomena.

Thus after a long historical evolution the intellectual positions have been reversed: whereas for the ancients ontological analysis [1] and ontological explanations absorbed everything,—even the sciences of phenomena themselves,—in a philosophical interpretation, now on the contrary empiriological analysis [1] absorbs everything and pretends to take the place of a philosophy of nature.

Physico-mathematical science is no longer mistaken for philosophy of nature as it was in the XVIIth century but it continues to take its place. At first it was confused with the philosophy of nature and then it displaced it.

There have been two consequences of this spiritual event, of this eclipse of the philosophy of nature, to the advantage of the sciences of nature: one consequence concerns science itself or the notion that is held of it, the other concerns metaphysics.

As for the consequences to science itself: this imperialism of the phenomenon of which I have been speaking reacted on the notion which was held of science, on the way in which science became aware of itself (among philosophers and also among scientists, with the help of philosophers). At one time the notion that was held of the science of phenomena (I am not talking about science in the Aristotelian sense here, but about the science of phenomena) was warped and forced, constructed according to a plan that was rigid and self-styled 'pure.' This was due to the fact that, taking the place of philosophy, it sought to set itself up as a counter-philosophy. So it had to do itself violence in order to exist not only for itself but in opposition to philosophy and *in the place of* philosophy, and bristled with means of defence and epistemological pretensions foreign to its nature in order to protect its position against philosophy's eventual offensive attack.

Thus arose the positivistic notion of science according to which science must keep itself undefiled, as from epistemological impurity, from every question and pretension about being, substance, cause, the "why," etc. Note that Kant, whose procedure was much more profound, more philosophical than Comte's did not seek to eliminate such notions from the sphere of scientific knowledge; he sought to phenomenalize them, which is not the same thing, to keep them the while he said: they have only a phenomenal value. Which, we repeat, is incorrect with reference to the theory of the understanding in the Kantian system, to the properly Kantian notion of phenomenon and to every possible use of the notions of cause, being, substance, finality etc. But it has its truth as regards the use of those notions in the circumscribed realm we are speaking about right now, that is precisely in the realm of the sciences of phenomena, in the sphere of the experimental sciences. It is not a sin of lèse-realism to phenomenalize notions if we consider their use precisely in the knowledge of phenomena as such.

Auguste Comte, on the contrary, purely and simply banished these notions from the domain of science and required that the scientist remain chastely distant from them while he constructs his knowledge of nature, because they are polluted by metaphysics. For positivism, science,—and every valuable knowledge,—is characterized above all by the elimination of every ontological preoccupation; that is the privilege of the positivistic age or state in opposition to the metaphysical and theological state. Every ontological preoccupation must be ruled out and with it every inclination to explain things by causes:

science is to be reduced to laws, to connections between phenomena; its sole task is to describe phenomena and to seek out the stable bonds between them, carefully substituting this notion of a well-established connection between phenomena for the notion of cause or *raison d'être*. This idea of science was not peculiar to A. Comte however. It was held by all his adherents among the scientists, particularly by the famous theorist of science and energistic philosopher, Mach.

Science must therefore abstract from the being of things and consider only the connections and relations which constitute the laws of phenomena. This is the sense in which Auguste Comte's famous formula can be best understood: that there is but one absolute principle and that is that everything is relative; for considered as a philosophical formula it is absurd. As Meyerson writes in describing this positivistic conception of science: "Even if we are called upon to formulate suppositions, hypotheses, these must have as their sole object an empirical law that is as yet unknown." [2] Auguste Comte himself wrote in his *Cours de Philosophie Positive*, "In order to be really judgeable every physical hypothesis must be exclusively concerned with the laws of phenomena and never with their mode of production" [3] (because to say mode of production is to say causality).

4. You can see how positivism introduced into the consciousness of the scientist a sort of holy dread of finality, causality and in the last analysis of intelligibility. We must not forget that this sort of ascesis, these positivistic

macerations imposed on the scientific intellect have been useful, have from a certain point of view provided a profitable discipline. What I mean to say is that they have countered the abuses due to the imagination, anthropomorphism, intuition,—the intellect. From a certain point of view positivistic science must, in order to keep itself pure, resist the intellect which proceeds too quickly to the explanatory cause and is not and never will be content to link one phenomenon to another,—a thankless task indeed! Under the impulsion of positivism science has tended to set itself up absolutely and divinely as a *pure discipline* of the phenomenon and its relations. Purity it was, yes, but purity that was likely to make it pure of reality too, and risked being the purity of emptiness and sterility. The danger is that this way of knowing ends up as a pure decomposition of the real into a dust of mathematical beings of reason (*entia rationis*), without ever grasping, wishing to grasp or even trying to reach, even in a quite indirect and obscure manner, the inner structures, the living treasure, the typical secret (the only one of prime importance for the mind), in short the unique, singular *name* of the various realities given to the senses. The scientist,—to tell the truth I am not talking about the scientist here, I am talking about the positivistic idea of the scientist: the scientist, as we shall see further on, does not behave in this way; but this is an ideal which the positivists have tried to impose upon him as an ascetic rule,—the positivistic scientist, the scientist as positivism conceives him, would wind up by analyzing the real perfectly in the quantitative and material order, yet on one condition: that he deal only with the corpses of reality.

This danger for science was clearly seen even before Comte's positivistic systematization, for it is inherent to the very handling of the physico-mathematical explanation. It was seen by Goethe when he had Mephistopheles say to the Scholar the celebrated lines: "Whoever wishes to know and describe the living, first seeks to eliminate the spirit; then does he hold the parts in his hand but what is missing, alas, is the spiritual bond."

> *Dann hat er die Teil in seiner Hand,*
> *Fehlt, leider! nur das geistige Band.*[4]

Such was reason's task as the XIXth century understood it. That is what positivistic materialism calls reason, reason as positivism imagines it and thinks to purify it in its scientific task.

Let us hasten to remark,—we shall come back to this later, that contemporary science, precisely while becoming more fully self-aware, is in full reaction against these positivistic pretensions and interpretations, against this false positivistic asceticism of the intellect. This reaction can everywhere be witnessed, in physics and in biology; but let us also hasten to add that, in order to lead to happy results, such scientific reactions must needs be based upon a sane philosophy. For Faust too was reacting against this materialistic conception of science; he too, in speaking of this merely material analysis of reality, cried out: "Skeletons of animals and bones of the dead." [5] But, it being his misfortune to live in an age that did not put Thomistic philosophy at his disposal to rectify this reaction, he concluded:

Drum hab' ich mich der Magie ergeben.[6]

"Therefore, now, I've given myself to magic." The danger of magic threatens every inordinate reaction against positivism and rationalism.

The Advent of Empiriological Thought and Metaphysics

5. We have been speaking of the first consequence of the advent of empiriological thought, the consequence that concerns science itself. There is another which affects metaphysics. We have just seen that this advent had as a first consequence the radical exclusion of the philosophy of nature: it was thought that there is not, there cannot be any philosophy of nature. This exclusion may be arrived at in very different ways: as far as pure, orthodox positivism is concerned there is no philosophy of nature for the very good reason that there is no philosophy at all: speculative philosophy consists only in reflecting on the sciences and coordinating them in an objective synthesis.[7]

But we need not restrict ourselves to the orthodox positivism of Auguste Comte, or to the even purer and more orthodox positivism of Littré. Even for modern philosophers who champion the rights of philosophy

alongside the scientific explanation of phenomena there
can be no philosophy of nature: because they admit the
positivistic conception of science and because they admit
that it exhausts the knowledge of sensible nature.

Well, what will be the results of all this for metaphysics? Clearly the advent of criticism and positivism
could not annihilate the mind's natural aspiration to first
philosophy. Metaphysics had to strive to put forth some
new branches. But under what conditions? The lesson
history gives us is singularly clear-cut.

After the failure of the great post-Kantian idealist systems wherein, we must not forget, a tremendous amount
of work on the philosophy of nature (the *Naturphilos-
ophie* of romanticism) was done in connection with work
on metaphysics and suffered the same fate; after the
failure of partial and timid attempts at speculative metaphysics founded on psychological introspection, in the
fashion of Victor Cousin and Maine de Biran, what do
we find? There is no more philosophy of nature, the entire field of knowledge of sensible nature is abandoned to
the sciences of phenomena, to empiriological knowledge.
The philosophers tried hard to set up a metaphysics, yes;
but being much more under the influence of positivism
than they realized they did not even dare conceive the
possibility of an ontology of sensible nature complementing empiriological knowledge. Well! once there is no
philosophy of nature, there is no speculative metaphysics.

There is only a reflexive metaphysics in which the

philosopher's task is not simply to reflect on the sciences in order to coordinate them into an objective synthesis, as Auguste Comte would have had it, but also to seek in the sciences, in the knowledge of phenomena as such, something undiscerned by the scientist but discernible by the philosopher. Let us distinguish here, two types of reflexive metaphysics. (There are others, we shall come back to these later.)

The first type is an idealist reflexive metaphysics: for example the simultaneously idealist and Spinozistic position of Brunschvicg. According to this position the mind tends to become self-aware in the course of history proportionately as it makes scientific progress. Progress in what science? Precisely in the science of phenomena. And it is this progressive self-awareness of the mind or the "spirit," immanent to the development of mathematical and physico-mathematical sciences, which constitutes both philosophy and spirituality. The error here lies not in seeking spirituality but in not wanting to seek where it comes from and, especially, in wanting to limit all spirituality to the mind's self-awareness in its scientific work. Thus the "spirit" appears as a formless and faceless god, a pure creative liberty without nature or essence which, in the worlds it is endlessly creating in order endlessly to surpass the old by the new, conjures up provisory and fleeting but always glorious images of its own abyss or rather of its own infinite void. The end-result of pursuing this philosophic tack is a sort of lay mysticism of Pythagoras' table and Foucault's pendulum.

Another type of reflexive metaphysics which is not
idealist but has, instead, a tendency to realism, is Bergson's. Within that physico-mathematical knowledge
which is rooted in the succession of phenomena but
ignores the reality of time and duration, he too is seeking,—but in another way,—for a metaphysical content
which, since it is sought therein, can evidently be found
only in time and duration itself. That is the way Bergson
arrives at his philosophy of duration. Note this well. I
believe it to be important for the understanding of Berg-
sonism: this philosophy of duration, this system of Bergson's, purports or rather purported (for Bergson later
sharply curtailed the metaphysical ambitions of his system) to be a metaphysics of modern science, a philosophy
or metaphysics of the experimental sciences. It represents
an attempt to seek the proper object of metaphysics, the
proper object of wisdom, in the entrails of the proper
object of science; not above, as metaphysical reason demands, but within the *formal* object of the experimental
sciences itself, as if physico-mathematics had a metaphysical content with which positive science is unwittingly pregnant. Bergson's intention is not to construct a
psychological philosophy but rather to embrace physics
so closely that he will discover at its heart a metaphysics
that is unknown to the physicist himself. The tendency
here is not idealist but realist,—a genuine effort to attain
to a real which is independent of the mind: an effort to
attain to this real not in being but in time, in pure change
which is the only metaphysical substance abstractable,

although in an illusory fashion, from the physico-mathe-matical web of phenomena.

From the fact that this is an effort to attain philo-sophically to the sensible real, this attempt approaches the philosophy of nature. It is an effort to penetrate philo-sophically (by means of intuition, which for Bergson is the reverse of scientific analysis) the realm of the natural sciences itself. So, from the noetic point of view, this does approach the philosophy of nature. Which is why Berg-son's ideas have in historical fact had an influence upon the life of the sciences, and on the ideas of many scien-tists, for example Hans Driesch and some English biologists.

But actually it is not yet a philosophy of nature; this conception remains a metaphysics, for its interest in the science of the physical world springs from the desire to find within it and by its means a metaphysical absolute which would be the absolutely last reality. To tell the truth, what this philosophy thinks it finds in this subsoil of physics whereto physics itself cannot penetrate, is something which it has itself placed there: a reality de-rived from psychological intuition and introspection. So that, while this pseudo-philosophy of nature tries to be a philosophy of physics, it nevertheless remains dependent upon the modern spiritualist tradition which began with Descartes and Leibniz and which seeks in introspection the means of transcending the mechanism of the natural sciences. In effect what we have here is a philosophy of nature which is really a metaphysics and an erroneous

one at that: for it affirms change as the sole reality while
it denies potentiality; affirms movement without anything mobile, change without anything that changes;—a
metaphysics which pretends to spring forth from science
under the intuitive eye of the philosopher.

SECTION 2 · MODERN REACTIONS AGAINST
THE POSITIVISTIC CONCEPTION OF SCIENCE

Pierre Duhem

6. The crises and the progress of science, the reflections of scientists and philosophers, inevitably shew forth
the unreality of the positivistic conception of science; it
is falling to pieces before our very eyes.

What have been the principal reactions against the
positivistic conception of science?

First I would like to discuss the reaction of Pierre
Duhem who is, as you know, as noteworthy a physicist
as he is an historian of the sciences. Duhem reacted very
strongly against the *second* of the consequences of the
advent of empiriological thought which we discussed in
the preceding section: the one which reverberated on
metaphysics and philosophy due to the exclusion of the
philosophy of nature. He wanted to show that there is a
place for a philosophy of nature, but he did so by carrying
to its extreme the positivistic conception of science, carrying to its limits the *first* of the two consequences we indicated, the one having to do with the notion of experimental science.

For him, physics and the science of phenomena in general has for its only object the pure mathematical legality of phenomena without any inquiry into causality. Physical theory is not an explanation nor does it seek in any degree to be an explanation; it is a system of mathematical propositions purporting to represent as simply and as completely as possible an ensemble of experimental laws. The scientist draws from the sensible world a certain number of observations and measurements and then, once these observations and measurements have been made, cuts every tie with the physical real. These observations and measurements are assumed, taken up into a pure work of Analysis whose highest and sole law is mathematical beauty. The result is a pure system of differential equations of which Duhem tried to give an example in his works on thermodynamics, a system without mechanistic meaning because it is without any properly physical meaning, and without any interest in physical causes or in the physical reconstruction of phenomena, without any physical imagery.

Then according to Duhem, once one is aware (as positivism is not) that this positivistic purification of science from every causal and physical pretension has a wholly mathematical character and meaning, it is easy to see that this same purification leaves room for another possible interpretation of nature, an interpretation wherein everything qualitative in the physical world would be restored. So Duhem's conception of science resulted in a sort of mathematical purism, and this reaction was very useful in this sense: that it showed forth the possibility of a qualitative interpretation of nature alongside physico-mathematical science. But his conception itself is open to criticism: Duhem fell into a conception of science, of the science of the physicists, that was too idealist, almost nominalist in character and at the same time,—from the point of view of the sciences themselves this is the most serious aspect in such a conception,—he suppressed the proper stimulation to physical research. Science became so pure in its mathematical symbolism that the principal and motivating appeal of physical *research*, namely the discovery of causes, the sense, the taste of the particular mystery to be discerned in physical existence, would have been completely lacking for the physicist had Duhem's conception of physical theory been correct. Today physics seems, on this point, to be in marked reaction against Duhem and his purely formal mathematicism.

Emile Meyerson and French Epistemology

7. A second reaction against the positivistic conception of science is Meyerson's, a philosopher of the sciences, not a physicist like Duhem. Meyerson did not take the metaphysical point of view in his studies; he took a purely and strictly epistemological point of view. His aim was to analyze the scientist's state of mind. Inquiring into the psychological and logical conditions required for the pursuit of science, he found that science is in actuality haunted by ontological and explanatory pre-

occupations. For example he showed in his book, "*L'ex-plication dans les sciences*," [8] that science requires the concept of thing, thing independent of the knowing mind. This word, thing, corresponds to one of the transcendentals recognized by the ancients: *res* was a transcendental. And science is so much in need of this concept of thing that it is constantly creating new, more or less fictitious things which it needs as principles of explanation.

In this work of Meyerson's may be found a host of interesting quotations showing that this interpretation is based upon the testimony of scientists themselves. Cournot, for example, wrote: "Whatever may be said in the modern scientific schools where any apparent lapse into metaphysics is feared above all else, mitigated as well as pure atomism implies the pretension of somehow grasping the essence of things and their inmost nature." [9] Meyerson, commenting on this passage, remarked: "All science rests upon the unconspicuous (since the nature of this foundation has been denied) but nevertheless solid bed-rock of the belief in being that is independent of consciousness." [10] So, in practice, science adheres to an implicit realism even when the scientist himself adheres to a metaphysical idealism. For it is absolutely impossible to imagine a scientific vocabulary exclusively made up of events and relations and excluding the notions of substance, cause, *raison d'être*, tendency, quality, force, energy or power (as we would say), potential state, actual state, etc. Actually science is constantly using such notions. And Meyerson is justified in saying that: "Genuine science, the only one we may know, is in no way and in none of its parts conformed to the positivistic scheme of things." [11] Which does not mean to say that this scheme has not had a real influence on science, but it has not succeeded in imposing itself on it in practice.

Science demands or presupposes concepts of philosophical or metaphysical origin (more or less altered and *recast* in transit, but that is another question); it seeks explanations of what has been observed, it tends invincibly to be explanatory in type, and that is important to remember. Meyerson rightly remarks: "it is not true that our intellect is satisfied with the simple description of a phenomenon however precise it be. Even if science is prepared to submit a phenomenon in all its details to empirical laws, it seeks more than that; it has always done so and continues to do so to this very day"; [12] it cannot help looking for "an explanation outside and beyond the law." [13] And this you see is flatly opposed to Duhem's interpretation, which limited science to the pursuit of mathematical legality.

Science must, then, be related in a certain way to real causes. Despite all the positivists had to say about it and despite what you will still find in many contemporary vulgarizations of science or scientific method, we are compelled to recognize that science does not escape the question "why." You remember how it has been dinned into our ears,—for it is one of the most famous commonplaces of our times,—that science is concerned with "how" and philosophy with "why." Well, to Meyerson's credit, he realized that the scientist, too, must deal with the question "why." He does not answer it in the same way as does the philosopher; he can happen to answer it in only the most rudimentary fashion. For many physicists,—at least as Eddington says, for those of the time of Queen Victoria,[14]—a phenomenon was not explained, its *why* was not given until a mechanical model of it could be constructed. Such was Lord Kelvin's position: "If I can make a mechanical model [representing the structure of matter, for example] I understand," said Lord Kelvin; "if I cannot do so, I do not understand." [15] But after all, this understanding which consists in constructing a mechanical model corresponds to a search for the "why"; the aim, however material it may be, is to understand.

In Meyerson's attempted restitution of ontological values two things are to be remarked: on the one hand there is a rudimentary philosophy implied as something *presupposed* by the very exercise of science; an implicit, unconscious philosophy which does not enter into the texture of scientific explanations but is present as presupposed. For example, the scientist is convinced, as we pointed out above, that things exist independently of the mind. Presupposed and postulated, this conviction does not enter into his science as a part thereof, but he needs it in order to construct his science. He is equally persuaded that it is possible for one's knowing powers to grasp things; that is, he is instinctively persuaded of the world's intelligibility, however ill defined it may be. And all that is philosophy.

On the other hand, the scientist has ontological and explanatory preoccupations which do enter into the texture of his scientific work itself. Thus there are relations to ontological reality that science needs not as presupposed but as integrating elements of its structure. When science elaborates upon the notion of electron or of quanta, it is dealing not with presupposed philosophical convictions but with properly scientific notions which enter into the very texture of science and, for this reason, have a certain explanatory value and bespeak a certain relation to ontological reality.

8. So from these two points of view, be it from the point of view of his larval or presupposed philosophy or from the point of view of the elements from the causal order which enter into his scientific explanations, the scientist owns to certain ontological interests. And yet,—a most remarkable fact,—being does not appear here except as an "irrational" element which science runs afoul of in its tendency to explain everything by reduction to the identical.

Here we come up against Meyerson's philosophy of science, which implies, especially at the beginning and afterward in a more qualified manner, the belief in what might be called the Eleatic functioning of thought: to explain is to identify; to explain two phenomena is to restore them to a superior identity. The mind's essential tendency in seeking the why of things is to eliminate every diversity as irrational. In that consists the natural and normal play of reason Eleatically or mechanically conceived. In short, as Meyerson says, everything must be explained by space.[16]

To this fundamental tendency of the human mind which is asserted without critical examination and confused with the natural exigencies of reason, is opposed the existence of what many contemporary authors, Meyerson included, call irrational elements; that is to say, elements that cannot be reduced to explanation defined in this way. These are elements which *resist* this need for identification. They form, as it were, "blind-alleys" forcing science to recognize such and such an element which provisorily or definitely cannot be submitted to this process of identification. These very numerous irrational elements,—tridimensionality of space, transitive action of one body upon another, the diversity of chemical elements, etc., have been most interestingly enumerated by Meyerson who has culled them from all the sciences, all the way from geometry and physics to biology.[17] In this way he has very well shown that even though science tend toward a mechanistic or rather a mathematicist explanation as its ideal (which is purely and simply true of the physico-mathematical sciences and true in some respect only of the other sciences), the mechanism or mathematicism in question is in any event only methodological; it could never be fully realized and, on the contrary, is constantly being thwarted and hindered for the good and the very progress of science.

This mathematicism is thus emptied of every dogmatic and philosophical pretension. "The scientist of today," writes Meyerson in another essay, "cannot say what the *essence* of the real is. This is what distinguishes his attitude from that of his materialist predecessor and even more from that of the mediaeval physicist; he no longer asserts that he can truly attain to the being of the real which, on the contrary, appears to him to be wrapped in the deepest mystery. Confronted by the real he feels as though he were in the presence of an enigma both admirable and troublesome: he contemplates it with an almost fearful respect which is not without some analogy to the emotion experienced by the believer before the mysteries of his faith."

Thus although science inevitably implies a relation to real causes and to being, it attains to them only under an *enigmatic* or mythical form. And yet, for Meyerson, this is the only way that we have at our disposal of attaining to the being of natural things. At least he does not ask himself if there is any other possible way of grasping the physical real. With the result that there is no other knowledge of nature than this enigmatic knowledge offered by the physico-mathematical and experimental sciences. It is highly remarkable that where the philosopher uses the term *intelligible being*, Meyerson and the contemporary philosophers of science use the term *irrational*. To express the same thing the former uses "intelligible being" where the latter uses "irrational" because being is grasped but blindly and enigmatically by the natural sciences and because, in fact if not in right, these philosophers recognize no other explanation of the physical real except that given by the sciences of nature. Being, which the idealist philosopher calls *intelligible being* because it holds no mystery for thought (for idealism only thought contains mystery, moreover it creates its object); being, which the Thomist philosopher calls *intelligible being* in a wholly different sense, because the inexhaustible mystery with which it abounds is precisely the substance of intelligibility, the dominating light of our thought;—being now appears as a reservoir of unintelligibility, a world of *irrational* elements.

So we may say that Meyerson's attitude is the inverse of Duhem's. Duhem reacted against the elimination of the philosophy of nature by pushing the positivistic conception of science to its extreme. Meyerson also reacted against the positivistic conception of science but he asserted that beside or, rather, above the science of phenomena there is no place for a philosophy of nature.

9. Close upon Meyerson's name it seems fitting to mention that of Professor G. Bachelard, another French philosopher who has done authoritative work in epistemology. On the characteristics proper to the contemporary scientific spirit, on its non-Cartesian character and on the very precise concrete relations which it sets up between experience and reason, you will find it very interesting to consult his work *Le Nouvel esprit scientifique*.[18] You will see that these relations, as he explains them, can be fitted very happily into the perspectives of Thomistic episte-

mology considered in its general noetic principles and in its particular theory of the *scientiae mediae*. See particularly Bachelard's important remarks on the "realizing function" of science which goes from the rational to the real, thinks out physical problems mathematically, and progresses by in some manner creating its object, the while it keeps in contact with the world and renews and deepens its thought by this contact with the object. Some explanation of these remarks can be found in our discussion, either in this present work or elsewhere,[19] of the nature of physico-mathematical knowledge and of the use it makes of *beings of reason* which have their foundation *in re* (*entia rationis cum fundamento in re*).

Bachelard's work very opportunely modifies Meyerson's and can serve as a useful precision of the latter's realism. However, lacking a regulative metaphysics, his work seems to incline overmuch toward idealism. I am convinced that only the Thomist theory of the *ens rationis* allows the *ideality* of the knowledge of nature to play its part (an immense part) without spilling over into idealism.

In our opinion Leon Brunschvicg is to be reproached for having jammed some very valuable epistemological views and analyses in among a most pernicious idealist metaphysics. There is no need to expatiate on his ideas here; an arbitrary interpretation too often deprives them of efficacy in the properly epistemological line. However, it would be unjust not to acknowledge in passing his important work in mathematical philosophy.

German Phenomenology

10. The third reactional movement I want to discuss with you is that of German phenomenology, which derives in great part from the work of Brentano, who had contact with Aristotelian philosophy. Its principal representatives are Husserl and especially Scheler. Max Scheler has had a great historical influence on a certain number of scientists, particularly biologists; for example, M. Hans André who is also a disciple of Father Gredt's and thus combines both Thomism and phenomenology. Phenomenological philosophy reacted *simultaneously* against both of the two consequences of the advent of empiriological thought we discussed above; against both the positivistic conception of science and the elimination of the philosophy of nature. So to this German phenomenological movement is linked an attempt to restore the philosophy of nature.

To put the matter briefly the phenomenologists oppose, to the attempt to "explain" in the mechanistic sense of this word "*erklären,*" the attempt to penetrate intuitively into the real itself, *verstehen,* to "understand." The aim is to construct an intuitive science revealing the essential articulations of the object; which object may be an object-phenomenon but is, in any event, an object offered to science and whose typical characteristics science seeks above all else to discover. Notions like *totality* (the whole explaining the parts instead of the parts explaining the whole), the *intuition of essences, typical* or *typological,* are given foremost importance. But behind this whole movement there is no regulative metaphysics capable of recognizing and setting the frontiers between scientific and philosophical explanation, so that there is great danger here of confusing their formal objects and not exactly of sacrificing the philosophy of nature to science (which was the positivist error) but, on the contrary, of making science itself into the philosophy of nature. That danger is peculiar to all vitalist or irrationalist reactions. These reactions are very useful in the measure to which they free thought from mechanism, but they are dangerous insofar as they open the door to irrationalism wherein analogy, for example, plays a wholly different role than it does in Thomistic metaphysics, and permits of explanations or so-called explanations that are more metaphorical than scientific in character. The laws (the style if one may so speak), proper to experimental analysis risk being confused with the style proper to philosophical analysis. Science can easily extricate itself from such a misunderstanding for, in the long run, it is always regulated and put right by experimental necessities, but philosophy is in danger of suffering from it.

Be it as it may with these three reactions we have been discussing, each of which has its advantages and its difficulties, and all of which are very interesting, it is evident that contemporary science is freeing itself of the positivistic conception of science.

III

Thomistic Positions on the Philosophy
of Nature

SECTION 1 · NECESSITY OF THE PHILOSOPHY
OF NATURE

1. As we have seen, to try to escape the problem of
the philosophy of nature is futile. We must face this
problem squarely and try to treat it doctrinally, for
itself.

In this first section I would like to show forth the
necessity of the philosophy of nature, the necessity of an
ontological or philosophical explanation of sensible
nature specifically distinct from that given by the experimental sciences, but complementing it. In scholastic language this section will answer the question *an est?* Does
and should a philosophy of nature exist?

Empiriological Analysis and Ontological Analysis

2. To begin this examination we must distinguish
between two ways of constructing concepts and analyzing
the sensible real. I propose to call these two types of

73

analysis by the following names: the one, the empiriological analysis; the other, the ontological analysis of
sensible reality.

When you observe any material object, that object is,
during your observation of it, as the meeting-place of two
kinds of knowledge: sense knowledge and intellectual
knowledge. You are in the presence of a sort of sensible
flux stabilized by an idea, by a concept: in other words
you are in the presence of an ontological or thinkable core
which is manifested by an ensemble of qualities perceived *hic et nunc*. I do not mean *thought* qualities but
sensed qualities, objects of actual perception and observation. If you come upon a plant during a botanical excur-
sion, you may ask yourself: what is a plant? and in that
case your interest lies in the direction of ontological
analysis. Or you may ask: how shall I classify this in my
herbarium? Here your interest is in another type of analysis: empiriological analysis.

There is, then, a twofold way of resolving our concepts (I am speaking of concepts that belong to the first
order of abstractive visualization) since their object is the
meeting-place of these two kinds of knowledge: sense-
knowledge and intellectual knowledge. For the sensible
real considered as such there is a resolution which may be
called ontological or *ascendant* toward intelligible being,
in which the sensible is always present and plays an indispensable role but does so indirectly: in putting itself in
the service of intelligible being and as connoted by it.
The other type of resolution is *descendant* towards the

sensible, toward the observable as such, insofar as it is observable. Not of course that the mind no longer refers to being, for that is quite impossible: being always remains; but here it enters the service of the sensible, of the observable and especially of the measurable. It becomes an unknown assuring the constancy of certain sensible determinations and measurements. Think on the one hand of the definition of a geosynclinal in geology, of verbal blindness in psychology, of a chemical species in chemistry, of mass or energy in physics; and on the other hand think of the philosophical definitions of the four causes, of transitive action and immanent action, of corporeal substance and operative powers. If you compare these two groups of definitions, you will find that they are arrived at by wholly different analyses and from different intellectual directions: in one case the definition is sought by means of possibilities of observation and measurement, by effectuable physical operations; in the other it is sought by means of ontological characteristics, of elements that constitute a nature or intelligible essence, however obscurely this essence may sometimes be attained.

So we are justified in distinguishing these two types of conceptual analysis and in saying that in one case we are dealing with an ontological analysis that is oriented toward intelligible being and in the other with an empiriological or spatio-temporal analysis oriented toward the observable and measurable as such. Furthermore, once we grasp the diversity of these two types of conceptual analysis, it also becomes clear that the same words, the same vocables can be used in the one case and the other, and be given entirely different meanings in each case. Think of the word *substance* for example, as the metaphysician uses it and again as the chemist or pharmacist uses it; there is almost no community of meaning between the two words: they are almost equivocal. Likewise for the word *property* as used by the philosopher who sees in a property the manifestation of an essence, and property as used by the experimental sciences. We must be forewarned of these differences in order to give the correct noetic coefficient to the words used by the scientist or by the philosopher.

3. Now that we have distinguished these two types of analysis and explanation: ontological and empiriological, I would like to make a few further remarks about them.

First Remark.—In speaking of empiriological analysis or explanation we have said that such analysis deals with real possibilities of observation and measurement, with effectuable physical operations. The permanent possibility of sensible verification and measurement plays the same role here as essence does for the philosopher. For the scientist the permanent possibility of observing and measuring is what the essence or quiddity is for the philosopher: it substitutes for it and takes its place. We have here, as it were, an effort running counter to the natural inclination of the intellect, because it involves turning back to the sensory act itself, to an operation to be performed in the sensible order, to an observation or a measurement, and viewing it as what is essential to and properly constitutive of the concept.

Whence it may be seen, as I have often pointed out, that the ascesis proper to the experimental sciences implies a certain opposition to the intellect, because in its natural movement the intellect starts from the senses in order to get to the intelligible, to the *raison d'être*, whereas here it is asked to turn back to the sense from which the concept derives, in order essentially to characterize this concept by means of operations performable by the sense under certain conditions.

Who understands this, understands the position of an Einstein, for example, in physics and the more apparent than real opposition between the philosopher and the scientist on questions such as time and simultaneity: such opposition disappears immediately because the type of definition is essentially different in the two cases. For the physicist who is aware of the epistemological necessities of his discipline, science seeks definitions not by means of essential ontological characteristics, not by a 'specific difference' that shows forth the essence, but by a certain number of physical operations to be performed under carefully defined conditions.

Second Remark.—We have said that empiriological analysis resolves the object into observable or measurable elements and thus goes from the observable to the observable, always remaining on the level of sensible operations, of obtainable observations or measurements. Now that statement needs further clarification. This analysis goes from the observable to the observable at least *indirectly*; for contemporary physical theories in microphysics and notably in the theory of the quanta, result in mathematical interpretations wherein the phenomena are no longer imaginable. For the imagination presents things to us as they appear in our scale of major dimensions, on a macroscopic scale, as possible subjects of a complete and continuous observation; whereas the scientist works in the atomic realm where, as Heisenberg has pointed out, the very possibility of a complete and continuous observation of phenomena vanishes. So he goes from a world of objects that can be represented by the imagination to a world of objects that cannot be imagined. This does not mean to say that this world is not a world of the observable: it always remains that; but these observabilities, if I may so speak, become discontinuous. The position of an electron can be precisely determined provided that its speed is not precisely determined, and its speed can be precisely determined provided that its position is undetermined. The fact remains that in each of these cases scientific analysis is dealing with a genuinely possible observation, but one which no longer implies the possibility of representing its object imaginatively. It is a kind of atomism of observation and mensuration which prevents the imagination from constructing for itself a model of the considered phenomenon, but we are always in the zone of the observable. Such a world is unimaginable by default or "privatively."

Third Remark.—The reason why we oppose empirio-logical analysis to ontological analysis is not because empiriological analysis abstracts from being; that is intellectually impossible and would mean falling into nominalism. Nor is it because empiriological analysis is valueless as far as reality is concerned: its aim is always to grasp reality; this analysis always has reference to being but its object is not to abstract the intelligible value of being for its own sake. Being is taken as the foundation of spatio-temporal representations and empirical definitions, or as the foundation of beings of reason [*entia rationis*] constructed by science and founded *in re*. Essence, substance, explanatory reasons, real causes, are attained in a certain way,—obliquely and blindly,—in substitutes that are myths or well-founded symbols, constructions of reason which the mind builds on the data of observation and measurement and from which it proceeds to meet things. And thus these primitively philosophical notions are, as we have already remarked, phenomenalized.

Having made these remarks in order to avoid possible misunderstandings, we may conclude that empiriological analysis deals with *sensible being but first and foremost as observable or measurable.*

4. We must also make some corresponding remarks about *ontological analysis* or explanation. Do not forget that we are at the first degree of abstractive visualization here: we are not speaking of metaphysics, we are speaking of the philosophy of nature and experimental sciences. This ontological analysis presupposes sensory activity;

not only does it presuppose it as does every human intellectual activity, but it remains in the limits of the sensible world; its object is finally characterized by means of experienced sensations. And yet,—this is an important, difficult and subtle point,—its object *precisely as intelligible* is not sensed; as intelligible (intelligible for us) it implies a reference to the senses but it is not sensed, it is not an object of observation. Take the notion which is nearest to sense experience, the notion of color, for example; as the object of a concept, as the object of an abstract *idea* (the idea of color) this object does not correspond to any physical operation to be done; it has reference to experienced sensations but insofar as it is an intelligible object it is not an object of sensation. Wherefore we may say that, in ontological analysis carried on at the first degree of abstractive visualization, being is considered in reference to sensible and observable data, but the mind consults this data in order to seek in it intelligible reasons that transcend the sense. That is why the mind, in acting this way, arrives at notions like that of color and, with even greater reason, at notions such as corporeal substance, quality, material or formal cause, operative power; notions which, although related to the observable world, do not designate objects which can by themselves be sensed or expressed in an image or in a spatio-temporal diagram. There is no possible image of *color* (which is neither white, red, green, nor any particular color). Such is the typical opposition between ontological and empiriological analysis.

Second Remark. For this reason we may say that in empiriological analysis we go from the observable to the observable, whereas in ontological analysis we go from the visible to the invisible, from the observable to the non-observable. We enter into a world which is not unimaginable privatively as is the world of microphysics, but unimaginable "negatively."

Third Remark. We must say a word about what might be called the paradox of these intelligible objects proper to the first degree of abstractive visualization: in themselves and as intelligible they are not the objects of a sensory act; my eye will never perceive the quality color as my intellect thinks it, nor will my imagination ever represent it. And nevertheless these objects humble the intellect in that data received from sense experience necessarily enter their definition,—a fact we must not forget,—so that there is an indirect but necessary reference to the sensory act in the case of concepts proper to the first degree of abstraction. Color as an intelligible object is not sensible; pure spirits as well as men have a concept of color, a notion of color: and yet they have not received it from the senses! But a man cannot understand the notion of color without referring to a sense experience.

The scholastics sought definitions which were apparently free of incommunicable sense experience; for example they defined white as *disgregativum visus*, that which diffuses sight. They were well aware however that a blind man has no idea of color; they constantly repeated the fact. But what they were trying to convey was that this idea, insofar as it is opposed to sense experience, designates a hidden *essence*, a quality, a kind of being which I can grasp only by referring to *my* experience and to the activity of *my* senses.

By way of parenthesis, that is the reason why Descartes hated these ideas. He hated the concepts proper to the first degree of abstraction and he refused them all objective value because they are not *pure* concepts, as he imagined mathematical concepts to be despite their link with the imagination. For him indeed, mathematical imaginability did not obstruct but rather assisted intelligibility. For him the notions proper to the first degree of abstraction cannot be of use to us to say what things are, can have no explanatory value (Cf. *Principia* I, 69, 70). He wanted physics to be intrinsically free from the senses; he demanded that it have a pure intelligibility which, as a matter of fact, lacked purity from the start because it was a geometric intelligibility. His way of making science specifically one was by brutally telescoping together the distinct and hierarchized noetic worlds which constitute it.

5. Thus even in ontological analysis there is, at the first degree of abstraction, an inevitable intrinsic reference to sensory operation. Nevertheless this analysis remains opposed to empiriological analysis on two points which we must note in passing.

First of all, the ontological type of analysis in the first order of abstractive visualization,—the analysis proper to the philosophy of nature,—honors sense perception more than does the empiriological type of analysis and expects more from it.

In ontological knowledge at the first degree of abstraction, sense-intuition is assumed into the mind's movement toward the intelligible; its value as knowledge, its *speculative* value enters into maximum play. When the philosopher deals with the humblest sensible reality, color for example, he does not do so by measuring a wave-length or an index of refraction, but by asking that the sight-experience to which he refers designate a certain nature, a certain quality whose specific intelligible structure is not revealed to him.

In so doing he respects the awareness peculiar to the senses; it brings him a content which as sensible is no doubt not intelligible, but which as sensible has a speculative value just the same. And it is thanks to this obscure speculative value which he respects in the sense, that he can turn the data furnished by the sense to the service of the imperfect intelligibility of an object of knowledge. The lived experience of the sense is respected for its own proper knowledge value, however inferior it may be.

In empiriological and especially in physico-mathematical analysis, on the contrary, it is highly noteworthy that the sense serves only to collect information which is furnished by instruments of observation and measurement; in so far as possible it is refused any knowledge-value properly so-called, any obscure seizure of the real. How could it be otherwise in that lifeless universe without soul or flesh, without qualitative depth, wherein abstract Quantity is mistaken for Nature? Descartes had his own good reasons for reducing sense perception to a mere subjective admonition, exclusively pragmatic.

For Aristotle, on the other hand, the act of seeing was the foremost example of the joy of knowing. Here you have two attitudes of mind which are fundamentally opposed from the outset and one may be permitted the observation that Aristotle's is the more human.

The true philosophy of nature honors the mystery of sensorial perception; it knows that such perception takes place only because the immense cosmos is activated by the first Cause whose motion passes through all physical activities so that, at the higher reach where matter awakens to *esse spirituale,* they may produce the effect of knowledge upon an animated organ. Wherefore the child and the poet are not wrong in thinking that in the starlight which comes to us across the ages, the Intellect who watches over us beckons to us from afar, from very far. It is highly instructive to observe here that the renaissance in Germany of the philosophy of nature due to the phenomenological movement, brought forth on the part of Mme. Hedwig Conrad-Martius, or of Plessner and Friedmann for example, a great effort to reinstate sense-knowledge. This is not the place to pass judgement on the particular effects of that effort. But to my mind, its very existence testifies to a fundamental, intrinsic need for the philosophy of nature which is too often neglected by modern scholastics.

The second characteristic in which ontological and empiriological analysis are opposed, lies as we have already said in the fact that the former seeks the essence above all else, an essence having a certain intelligible constitution. I do not know this essence *in itself*; I cannot know of color what an angel knows of it. An angel not only has the idea of color but, by this idea, he knows what color is; he knows the essence of this quality. I have the—human, not angelic—idea, I have the concept, the abstract notion of this same quality; by this concept I grasp an intelligible essence. I cannot say in what it consists and in order to say what it consists in, I am forced to refer humbly to my sense experience. Nevertheless it is in this intelligible essence itself that I am especially interested; that is what my concept seeks to attain. And this is where the ontological notion is opposed to the empiriological notion, which does not primarily designate an intelligible essence, but especially designates concrete possibilities of observation and measurement. For ontological analysis, sensible data are mere, albeit indispensable means, a means of designating the essence; they are not the essential element of the definition and of the notion as they are for empiriological analysis.

6. The fact remains that, despite this basic difference of orientation, ontological analysis at the first degree of abstraction *cannot be disengaged* from the sensible given; it definitely rests upon it. This is true even for the highest concepts in the order of the first degree of abstractive visualization. I am insisting on this because I think that

there are some important points here that are not always sufficiently well brought out; which explains why there is a tendency to confuse the philosophy of nature and metaphysics.

Compare some concepts borrowed from these two wisdoms: from the philosophy of nature which is a particular wisdom and from metaphysics which is wisdom purely and simply in the natural order. Take notions like those of form and matter, body and soul,—I am intentionally choosing the highest, most philosophical concepts which belong properly to the philosophy of nature. Of themselves and primordially these are not metaphysical concepts; they belong to the philosophy of nature. Compare these concepts with properly metaphysical concepts such as those of potency and act, essence and existence. There are signs that tell us clearly enough that these concepts belong to different intelligible degrees, for potency and act, essence and existence are found in purely immaterial beings such as "separated forms": in a pure spirit there is act and potency and the distinction between essence and existence; but in the realm of pure spirits there is neither prime matter nor substantial form, nor body nor soul. Well, does that mean that there is a simple difference of topographical distribution between these concepts, or is there a difference in intelligibility itself? In both cases the mind in its work of conceptualization, in the formation of concepts and definitions, tends toward intelligible being, seeks to grasp intelligible being; it does not fall back on the senses, as it does in the empiriological analysis we have been speaking of. Yet between the concepts of form and matter, soul and body, and the concepts of potency and act, essence and existence, there is a real *difference of intelligibility*; the degree of intelligibility of these concepts is not the same.

In both cases the senses are at the source of our knowledge. It is clear that all our ideas come from the senses; none escapes this law; but in the case of the concepts proper to the philosopher of nature, the sensible remains irremediably bound to the concept itself. That is what differentiates intelligibility at this degree from metaphysical intelligibility. The abstract notion of soul cannot be conceived without the notion of body; these are correlative notions, since the soul is the substantial form of the body. And we cannot conceive the notion of body without that of organism,—*caro et ossa,*—we cannot conceive the notion of organism without that of qualitative heterogeneity, nor the notion of qualitative heterogeneity without that of sensible properties; and we come finally to color, resistance, hardness, etc., which we cannot define except by an appeal to sense experience. So ultimately this notion of the soul, which is the most philosophical, the most ontological, the highest in the order of the philosophy of nature, cannot be conceived without this reference to sense experience, I mean in the very understanding of the notion itself. Whereas the notion of essence or of existence does not imply this reference to sense experience in its very definition, in the elements which integrate its definition: it refers to sense experience as to an analogical paradigm. There is an analogy between the perception which the sense has (in its fashion) of the existence of a *res sensibilis visibilis* and the intelligible value presented by the notion of existence, but this is a simple analogical relation; there is no reference to the sense in the very constitution of the notion itself.

The same is true of the notion of form, which is not conceived without that of matter; form and matter are not conceived without the notion of body and the notion of body leads us back finally to experimental elements.

A while ago we remarked that, from our present point of view, Descartes seems to have wanted above all else to make of the science of sensible nature a science intrinsically free of the senses. In other words, he wished to raise physics (since for Descartes there is but a single Science, specifically one) to the same degree of intelligibility as Mathematics or Metaphysics itself, so that the notions used therein would not imply this humiliating and necessary reference to sense knowledge. You see here how serious it is for a metaphysics to refuse to the senses any value with respect to speculative knowledge, and to accord them only a purely pragmatic or affective value, as did Descartes.

Let us conclude our remarks about ontological analysis by saying that *ontological analysis at the first degree of abstraction deals with sensible being but deals with it first and foremost as intelligible.*

The Philosophy of Nature Differs Specifically from the Natural Sciences

7. *Two specifically different types of knowledge* correspond to the two types of analysis or explanation we have just distinguished. There is a specific difference between the knowledge which uses empiriological analysis, the empiriological mode of defining, and the knowledge which uses ontological analysis, the ontological mode of defining. This specific difference between experimental science and the philosophy of nature was disregarded by the ancients.

What is the ultimate principle for the specification of the sciences? Thomistic logicians answer that it is the mode of defining, *modus definiendi.*

The ultimate principle for the specification of the sciences is taken in fact, not from the *terminus a quo* of the abstractive operation but from its *terminus ad quem.* The abstractive operation considered in the typical ways in which it withdraws from matter (*terminus a quo*), provides the foundation for the three *generic* orders of abstraction. Considered according to the typical ways in which it constitutes the object at a certain determined degree of immateriality or knowability (*terminus ad quem*), it founds the specific diversities between the sciences; and these diversities can be found within a same generic order of abstraction.[1]

Thus the orders or degrees of abstractive visualization of which we have been speaking,—first, second and third degrees of abstraction,—correspond to *generic* differences in knowledge, and these generic differences are determined by the typical ways in which the mind, in the operation of abstractive visualization, withdraws from matter (*terminus a quo*) and leaves it behind.

But there can be *specific* differences between sciences which belong to the same generic degree; for the ancients, for example, geometry and arithmetic were two different scientific species. Descartes and all modern Mathematics after him tried to make a single science out of them, but for the ancients they were two different scientific species both of which were nevertheless at the mathematical degree of abstraction and both abstracting from sensible matter. How then was one distinguished from the other? By reason of the typical ways in which the mind, in abstractive visualization, not only withdraws from matter but positively constitutes, sets before itself things at a certain determined level of objective immateriality and intelligibility (*terminus ad quem*).

8. And in what does this *terminus ad quem*, this principle for the specific differentiation of the sciences, ultimately consist? It consists in the *modus definiendi*, in the typical fashion of conceptualizing the object and of constructing notions and definitions. ". . . licet in una scientia tractentur diversae res seu quidditates, quae in se possunt habere diversam perfectionem et diversam abstractionem, sicut Metaphysica quando tractat de Deo et de praedicamentis, Physica quando tractat de elemento vel de anima, tamen *semper est idem modus definiendi*,

quia sicut elementum definitur ut mobile, ita anima ut actus rei mobilis, et sicut praedicamenta tractantur ut participant rationem entis, ita Deus ut prima causa totius entis, quod est sub eodem ordine omnia tractare, sicut in eodem corpore sunt diversae partes habentes diversas perfectiones, omnes tamen conveniunt in ratione informati ab eadem anima . . . Definitio ut tali modo abstractionis facta, est ratio formalis sub qua respectu conclusionis, quae per illam illuminatur." [2]

Because of this doctrine the Thomistic logicians consider that the different parts of the philosophy of nature constitute one single specific branch of knowledge,—precisely because the *mode of definition* is of the same type throughout and has to do throughout with being itself as mobile or mutable: ". . . formalis ratio entis mobilis adunat omnia quae tractat Physica sub unica ratione mobilitatis." [3] And the reason why St. Thomas seems to set the philosophy of nature and the sciences of nature in one same specific class wherein the different degrees of concretion of the object only entail differences of more or less,[4] is precisely because at his time the natural sciences, except for certain already mathematicized realms like those of astronomy and optics, had not yet won their methodological autonomy and still constructed their definitions according to the same typical pattern as the philosophy of nature. In all these cases the *modus definiendi*, the mode of conceptualizing the object, the type of notional analysis was the same.

Nevertheless John of St. Thomas pointed out that in

the generic sphere of the first order of abstraction there may be specific differences, for example between *philosophia naturalis* and medicine; for, he said, although both of them abstract from individual matter and not from sensible matter, yet medicine's object is more concrete,— the body considered as a thing to be healed,—than is the mobile body as such; "magis *concernit* materiam corpus ut sanandum quam corpus mobile ut sic." [5] Although medicine is at the same generic degree of abstraction it is at a specifically more concrete degree than is natural philosophy; here we have a same generic degree as to the way we *withdraw* from matter, but not the same specific degree as to the term toward which we *tend*, which term is made known by the definition, by the mode of defining.

9. If this is the case, if the ultimate principle of specification for the different kinds of knowledge is the mode of defining or the way in which notions are constructed, then it is evident that in the generic sphere of intelligibility of the first order of abstraction, the notions and definitions resulting either from empiriological analysis, wherein all is resolved into the observable, or from ontological analysis wherein everything is resolved into intelligible being, correspond to specifically distinct kinds of knowledge.

The way in which the sciences of nature,—simple experimental sciences or physico-mathematical sciences,— conceptualize their object and construct their definitions is typically different from the mode of defining and conceptualizing proper to philosophy. The conceptual vocabulary of the natural sciences and that of the ontological interpretation of nature are typically different. (What we are calling conceptual or notional vocabularies correspond to what the ancients called *ratio formalis sub qua*, a technical expression we shall be using later on.) Even if the philosophy of nature and the sciences of nature use the same words, the mental word or concept signified by the same word is formed in a typically different way in the two cases.

The Philosophy of Nature and the Natural Sciences Are Mutually Complementary

10. *Let us say then that there are experimental sciences of phenomena specifically distinct from the philosophy of nature, and that there is and must be a philosophy of nature specifically distinct from the sciences of phenomena.* For, as Meyerson has shown, the experimental sciences do actually imply an ontological tendency and reference,—which they do not and cannot satisfy. These sciences aim at being (as real) but, at the same time, distrust it (as intelligible) and turn back to sensible phenomena. As we have already remarked, they must in a certain sense go counter to the intellect's inclination in order to constitute themselves according to their pure epistemological type.

The sciences of phenomena thus testify that nature is knowable and that they only know it in an essentially unsatisfactory manner.

Accordingly *they ask to be complemented by another*

knowledge of tue same sensible universe, an ontological knowledge in other words, the philosophy of nature. Not only are we saying that the sciences deepen and quicken the intellect's desire for more profound or higher truths, just as the philosophy of nature itself quickens the intellect's desire for metaphysics, but we are also saying that as knowledge ordered to a certain term, the experimental sciences need to be complemented. Not, of course, that they need to be complemented as to their own rule of explanation nor as to the formal object that specifies them, but as to the terminus at which they aim, which is the sensible real. Insofar as this terminus offers to the mind certain riches of reality, a certain density of knowable reality, insofar as it has a certain intelligibility-appeal, as we shall say further on, briefly insofar as it is mutable and corruptible, this terminus is known in an essentially insufficient manner by the sole help of the proper vocabulary, the objective grammar of empiriological knowledge. Therefore, this knowledge needs to be complemented by another which, being likewise at the first degree of abstractive visualization, will attain to the very intelligibility of the mutable and corruptible real.

11. The experimental sciences ask to be complemented by the philosophy of nature; but the inverse is equally true: *the philosophy of nature asks to be complemented by the experimental sciences.* By itself it does not give us a complete knowledge of the object in which it terminates, that is, of sensible nature.

We are now in a position to understand the characteristics of this philosophical knowledge, of this ontological analysis of the sensible real, by opposing these characteristics to those of empiriological knowledge. The philosophy of nature resolves its concepts into intelligible being itself: it makes use of an ontological type of analysis which is open to the natural movement of the speculative intellect and it seeks to attain to the essence of things. It depends upon experience in a more cogent manner than does metaphysics; it must be able to submit its judgements to sense-verification because it belongs to the first order of abstractive visualization: but it is nevertheless a deductive knowledge, assigning principles of being and intelligible necessities in the same measure to which it has grasped the constitution of its objects. It properly belongs to such scientific knowledge to instruct us on the nature of the continuous, of number, quantity, space, movement, time, corporeal substance, transitive action, vegetative and sensitive life. It may even consider the relations of the universe to its first cause, as Aristotle does at the end of the *Physics*.[6] But because of its very structure, this ontological type of knowledge must forego explaining the detail of phenomena, exploiting the riches of natural phenomena:—an important point which was not at all clear to the ancients.

It may certainly be said that, from this point of view, the great modern scientific movement since Galileo freed philosophy and ontological knowledge from a host of tasks which it had assumed and which really did not belong to it. The explanation of the detail of phenomena belongs to science, to empiriological knowledge and analysis, whereas the philosophy of nature is already a wisdom. It is not purely and simply wisdom as metaphysics is; it is wisdom *secundum quid*, relative and inferior wisdom; wisdom in a given order because it is concerned with truly first principles, but wisdom in a particular order because it is concerned with the first principles of sensible nature. Now all wisdom is magnanimous, unconcerned with the material detail of things; so in this sense it is poor, and free as are all the truly magnanimous. And this wisdom which is the philosophy of nature is *obliged* to be poor; it must be resigned, indeed it should be honored to know the real with poor or humble means, without pretending to exhaust the detail of phenomena or to count the pebbles in the stream.

It must be borne clearly in mind that the essence of sensible things remains for the most part hidden to us, I mean in its ultimate specific determinations; below man and human things sensible realities do not reveal themselves to us in their specificity. We can have an essential knowledge of certain very general objects such as those we have already mentioned: vegetative life as opposed to sensitive life, life as opposed to inanimate matter, etc., but these are extremely general realities. When we want to arrive at the specific distinctions and diversities of things, we cannot discover the essence. Our understanding here is blind and must proceed by signs.

That is why there is no other science, no other knowledge of natural phenomena than empiriological science (which proceeds by signs) that is humbly content to explain in terms of the observable without seeking to discover the essence. It bears on the essence, but blindly, without discovering it; it bears on ultimate specific determinations but without discovering them in themselves. This science is not philosophy! And yet the philosophy of nature needs it; it demands this non-philosophical knowledge in order that the object in which it terminates be attained with sufficient completeness. For the object in which it terminates is sensible reality and sensible reality is not made up only of general objects like those we have been talking about: space, time, life, corporeal substance, etc.; it takes in the whole specific diversity of things.

So, as science, as knowledge, the philosophy of nature asks to be complemented by the experimental sciences, by empiriological knowledge which nevertheless differs specifically from it. And this clearly indicates that the philosophy of nature and the experimental sciences belong to a same generic sphere of knowledge, that they are both related, for very different reasons, to the first order of abstractive visualization; it clearly indicates that the philosophy of nature is fundamentally distinct from metaphysics. Metaphysics has no need to be complemented by the sciences of phenomena; it dominates them, is free of them.

But the philosophy of nature demands to be complemented by these sciences because they are two species of the same epistemological genus: both of them belong to the same order of abstraction (this is true of physico-mathematical sciences at least as to their matter and term for, as we know, the ruling principles of explanation of these sciences are not physical but mathematical). Which, as we pointed out above, is the reason why the philosophy of nature is much more narrowly and con-strainedly dependent upon experience than is meta-physics, which also derives from sense experience but, unlike the philosophy of nature, needs not verify its judgements therein.

Answer to a Difficulty

12. What we are saying is that there is a *specific* difference, a distinction of essence, between the philoso-phy of nature and the natural sciences. And we have seen the reasons for this,—reasons which are to my mind wholly decisive.

Now to some minds, accustomed to the scholastic view-point of the XVIth and XVIIth centuries, this solution may present some difficulties. They will will-ingly agree that the natural sciences make up a particular noetic universe if taken *historically*, as they are consid-ered *in fact* by scientists themselves. They agree that in the scientists' own opinion, in the way in which they conceive, conduct and advance their own science, keep-ing it strictly separate in its proper development as sci-ence from philosophy and philosophical problems (even though personally they may possibly be interested in philosophy and may have worked out a general conception of the world, as have several eminent physicists in our own day),—they agree that taken in this way the natural sciences do appear to be specifically distinct from the philosophy of nature.

But their difficulty lies in granting as much to the natural sciences considered *in themselves*. It seems to them, for example, that experimental psychology is only the *inductive basis* of rational psychology and that, being but the inductive part of a science, it is not a complete science by itself. Since they do not by themselves constitute a complete science, the experimental sciences of nature do not constitute a separate scientific species; consequently they are not specifically distinct from the philosophy of nature and are only its inductive basis.

13. Our answer to this is that a science may be specifically determined without necessarily and for that reason constituting a complete type of knowledge by itself. On the contrary, we have just been insisting that the philosophy of nature and the natural sciences need each other for their mutual completion. To our mind they are in a relation analogous (metaphorical analogy) to that of the soul and body; not inasmuch as the body and soul constitute a substantial whole, which has no sense in the epistemological order,—on this point our comparison is deficient; but inasmuch as the integrity of the reality-to-be-known, of the real term to which knowledge tends at the first degree of abstractive visualization, requires that philosophy and science be complemented one by the other, and inasmuch as, on the other hand,

the body and soul differ from each other not only in
degree but in nature or essentially.

It is precisely because of the inadequacy of all abstractive knowledge in respect to inexhaustible reality
that the human sciences are distributed according to
generic or specific degrees enveloping in their amplitude
a whole multiplicity of things or realities which differ
specifically one from the other.[7] And this same inadequacy explains the fact that, with respect to the same
real term to be known, sciences which are specifically
diverse by reason of their typical mode of defining are
essentially called upon to complement each other, and
that therefore each of them should be considered as incomplete; not, indeed with regard to its ultimate specifying object but with regard to the term wherein its judgements are verified. We have already remarked that, for
the ancients, the sciences of number and of the continuous were specifically distinct from each other. And yet,
so true is it that they ask naturally to be complemented
by one another, that modern mathematics has tried and
is still trying,—without any real success in our opinion,—
to make one specifically same science of these two parts
and to embrace both of them within a same *ratio formalis
sub qua* referring to the science of number. Turning to
the practical order, we see likewise that moral philosophy
and prudence answer to two specifically distinct types of
knowledge and yet it is evident that with respect to the
term toward which they tend,—the regulation of human
conduct, these two specifically distinct types of knowl-

edge, the latter of which is not even a science, have an internal need to be complemented by one another.

Experimental psychology is not merely the inductive basis of rational psychology; it constitutes by itself a science of the empirical order and differs specifically from the ontological knowledge of the soul. Experimental psychology asks to be complemented by this latter knowledge, just as this ontological knowledge asks to be complemented by experimental psychology. The more closely it approaches its pure epistemological type, the more it appears as something other than a simple inductive basis for rational psychology; for it obeys another rule of conceptualization, another *modus definiendi*. And rational psychology can derive philosophically useful facts and data from the scientific data collected by experimental psychology only on condition that it make a philosophical exegesis of these materials, illumine them by principles and philosophical lights which experimental psychology itself does not know. Generally speaking, it would be extremely imprudent for the philosophy of nature to look upon the natural sciences as a simple inductive basis for its own researches; but this is especially true with reference to those natural sciences wherein the real is treated mathematically and is conceptualized (at least in the most theoretical parts of such sciences) into beings of reason which are founded in the real. By so doing, the philosophy of nature would risk falling short of its own inner law and betraying the truths to whose pursuit it is ordered. For we have here two typically different universes of intelligibility, albeit belonging to one same generic degree of abstraction. In any event, the fact that the experimental sciences of nature do not by themselves constitute a complete branch of scientific knowledge,—I mean with respect to the real term within which their ultimate formal object is differentiated,—does not in any way oblige us to refuse them the right to constitute a separate scientific species.

The Subordination of the Empiriological Realm to Mathematics or to the Philosophy of Nature

14. We have now to investigate some more particular considerations concerning the empiriological realm.

Looking at this domain we can see that it is necessarily subject to the *double attraction* of mathematics and philosophy. For in actual fact the resolution of concepts into the observable and the measurable as such is not sufficient. Empiriology needs to be linked to a deductive science because in deduction is found the most perfect type of scientific explanation, and the deductive science to which empiriological knowledge is linked plays a formal and directive rôle with respect to experience. In more precise scholastic terms, empiriological analysis as such is either properly or improperly *subalternated* to a deductive science. And there are but two deductive sciences of a pure type: mathematics and philosophy.

What does this word, subalternation, mean? In John of St. Thomas' treatise on *Logic*, q. 26, art. 2, there is a very complete treatment of the theory of subalternation.

A science is said to be subalternated to another when it derives *its principles* from this other science, which is called the subalternant. The subalternate science does not by itself resolve its conclusions into the first principles of reason, into self-evident principles, but the subalternant science resolves its own conclusions into first principles and these conclusions of the subalternant serve as principles for the subalternate science. To use the classic example given by the ancients, geometry is the subalternant science with respect to optics (subalternate science) which explains the properties of light rays by geometrical laws. In this example of optics and geometry there is, to use the terminology of the ancients, *subalternation as to principles* because there is *subalternation as to the subject* of these sciences. The object or subject, (either of these words may be used here), of the subalternate science adds a difference which is accidental with respect to the object of the subalternant science. Thus acoustics is a subalternate science to arithmetic. Its object, says John of St. Thomas, is number, the object of arithmetic: but to that object it adds this accidental difference: *sounding* number. Optics is a subalternate science to geometry: its object is visual line, *linea visualis*; *visual* is an accident added to the object *line*, the proper object of geometry.[8]

These examples have to do with that type of empiriological analysis which is subject to mathematical interpretation. We may give a name to this sort of empiriological analysis in which the sensible is interpreted math-

ematically; we may call it *empiriometrical analysis*. Here
we are dealing with a sensible given which is attracted
by mathematical explanation and is not only attracted to
it but is drawn into the mathematical sphere of intelligibility, subject to its properly mathematical rules of interpretation and intelligibility, and thus integral with it.
In other words we are dealing simultaneously with a
science which is *subalternate* to another and with an *intermediary* science. We have here two things: subalternation and *scientia media*; astronomy is subalternate to
mathematics and is at the same time an intermediary
science. We may note in passing that the examples given
by St. Thomas of the sciences subalternated to mathematics, *musica, perspectiva, astrologia,* acoustics, geometric optics and astronomy are at the same time examples of *scientiae mediae,* materially physical and formally mathematical. This must necessarily be so in such
cases; the science subalternated as to its object must necessarily be an intermediary science, formally dependent
on the order or degree of the subalternant science, since
it considers and explains its own proper object (sounding number in acoustics, for example) only insofar as it
connotes the object of the subalternant science: number.
In other words, it considers and explains its own object
only insofar as this object comes within the formal
reason (*ratio*) or rule of explanation of the subalternant
science. So we are dealing here with a subalternate science which is also an intermediary science, *scientia
media,* and simultaneously belongs to the physical degree

of abstraction, materially, and to the mathematical degree formally.

From this you can see that the mathematical sciences tend, if I may so speak, to ravish the philosophy of nature of its proper object. The mathematical sciences, being deductive and explanatory sciences draw the sensible real into their proper sphere in order to explain it and consequently to construct a system of explanatory reasons and causes which takes in all the sensible real and explains it, not by ontological causes and principles which are *entia realia* of the intelligible order, but by mathematical beings of reason (*entia rationis*) constructed for this purpose, due respect being given both to experimental and numerical data gathered in the world of nature and to rules of mathematical computation and systematization. So there is a constant coming and going from observed and measured real beings to mathematical beings of reason and vice versa. And the more the mathematical ensemble thus wrought becomes full, rigorous and able to explain a great number of phenomena with a small number of principles, the more perfect the explanation will be. It will be perfect but, to explain the sensible real, it will use mathematically constructed entities and the danger will be great,—not inescapable but great,—of mistaking these mathematically constructed entities, *entia rationis* with their foundation in reality, for ontological causes, for *entia realia* explaining the essence of the physical real.

You can see how the intermediary sciences we have

been speaking of, science of the empiriometrical types, will tend to what might be called a mechanistic pseudo-ontology: ontology because in "beings of reason" there is "being," and "pseudo" because a being of reason is not a real being. This pseudo-ontology is mechanistic, let us say more generally, "mathematicist," for at certain times, as in our day, these sciences seem to veer toward a more Pythagorean than geometric or Cartesian style of explanation.

As a matter of fact, however, this mathematicist pseudo-ontology has of itself only a methodological value for the science which tends toward it. To the measure in which the science in question is formally mathematical, a science in which mathematical entities and mathematical principles of deduction play an essential or "constitutive" role, to that same measure will there be a necessary tendency to the Pythagorean or mechanistic ideal, without there being for all that the slightest necessity of tending to a philosophical or properly ontological mathematicism. The methodological mathematicism to which this tendency necessarily leads will consist of a system of explanation permitting the ensemble of the observable real to be deduced by means of ideal entities founded in the real. This mathematicist explanation inevitably runs up against a very considerable residue of irrational elements, but it also tends to reduce these irrational elements as far as possible.

So what we have in this instance is a material and quantitative analysis of sensible nature which seeks to reconstruct phenomena in a closed world which is a substitute for first philosophy (the world of mathematicism, but of a merely methodological mathematicism which can but be taken erroneously for a properly ontological and philosophical mathematicism). Epiriometrical analysis tends toward this world of mathematicism as toward its asymptote and the danger is that, before arriving at ontology properly so-called, (philosophical ontology) the mind may stop at this pseudo-ontology built of beings of reason and constituting a closed universe.

15. We have been speaking of what may be called empiriometrical analysis of natural phenomena. Besides this type there is another wherein concepts are resolved into the observable but without being subject to the mathematical rule of explanation. Here we are dealing with concepts defined in terms of certain possible external or internal observations but not essentially dependent upon a mathematical interpretation and deduction of the sensible real. Since this analysis constructs "schemas," as it were, which include a certain number of sensible determinations and empirical characteristics, we may call it *empirioschematic* analysis.

Insofar as it escapes the attraction of mathematics, so will it be attracted by another deductive science, this time philosophy: the philosophy of nature and beyond that, metaphysics. It tends, not toward the pseudo-ontology of mathematicism, constructed of ideal entities and causes but toward true philosophical ontology constructed of real causes and principles. It will be attracted to this type of deductive explanation but, note this well, it must always remain distinct from it. This "typological" experimental analysis, to borrow Hans André's expression, is ruled by the heaven of philosophy but remains on earth. Here, too, there is subalternation (or rather, subordination or infra-position since the subalternation here is improperly so-called), but of an entirely different mode from the one we have just discussed: here we have 1.—subalternation without the constituting of a *scientia media*, an intermediary science; 2.—subalternation improperly so-called.

1. In empiriometrical analysis, for example in astronomy or optics, we had a materially physical and formally mathematical science; but here we have, for example in typological biology, a science resolving its concepts into the observable, oriented toward philosophy, but not formally philosophical in the way in which astronomy is formally mathematical. This science is not astride both experience and philosophy as astronomy is astride sensible observation and mathematics. We may note here that generally, and even in subalternation properly so-called, there may be subalternation without the formation of an intermediary science. In scholastic terms, there may be subalternation not as to principles and *as to the object*, but *as to principles only*.

This is another type of subalternation recognized by the ancients; it is due solely to the fact that the subalternate science's means of demonstration depend on principles which it receives from another science, but the object of the subalternate science does not add any element that is new, or from a different order, to the object of the subalternant science. The example given by the ancients of this subalternation properly so-called and by reason of principles only, was the example of theology, which has the same object as the intuitive knowledge of the blessed but is nevertheless subalternated to it as to its principles, which it receives from this superior science by the intermediary of faith. Such a subalternation is possible only where the subalternate science attains to the same object as the subalternant science under a diminished light. In such a case the subalternate science belongs as to its formal reason to a degree of specification inferior to that of the subalternant science and cannot constitute a *scientia media* with it.

In other words the subalternant science and the sub-alternate science both bear on the same *thing*, presenting itself with the same intelligibility-appeal. [I am suggesting that we translate what the ancients called *ratio formalis objecti* UT RES or *ratio formalis* QUAE by *intelligibility-appeal*.] Here, in the case of the philosophy of nature and of the empirioschematic experimental sciences (which, as we shall see immediately, are infra-posited rather than subalternated to the philosophy of nature) the intelligibility-appeal in question is the mutability of nature, its mobility, its characteristic of being sense-perceivable. The philosophy of nature and the experimental sciences bear on the same existing thing; the sensible real as mutable, but their rule of conceptualization and explanation, what may be called their objective light, is different since, in one instance it is empiriological, in the other ontological. This objective light corresponds to what the ancients called *ratio formalis objecti* UT OBJECTUM or, again, *ratio formalis* SUB QUA.

2. The subalternation of the empirio-schematic sciences to the philosophy of nature is not subalternation properly so-called, as is that of theology to the knowledge of the blessed, and of optics to geometry. This subalternation is improperly so-called and should simply be designated by the more general expression of subordination or infra-position. The reason for this is that the experimental sciences are specified by an autonomous type of analysis and notional vocabulary, distinct from the very outset from those of philosophy. Because of the empiriological character of their definitions and notions, these sciences do not use the conclusions of philosophy as principles for their own demonstrations. Don't let the word, subordination, lead you to think that! That would be an absurdity since what we have here are conceptual vocabularies which are foreign to one another and opposed in their fundamental directions; and since the experimental sciences by themselves are in direct contact with natural evidence and do not need the intermediation of philosophy. But *when* the experimental sciences want fully to resolve their object in the light of first intelligible principles, then they must have recourse to philosophy in order to be complemented by it: that is to

say, they must give place to philosophy. Furthermore, the empirio-schematic sciences themselves *need* the conclusions and truths established by the philosophy of nature, *not indeed as constitutive principles* but, to use a Kantian distinction, as *regulative principles*, as directive principles orienting thought and research, but not entering into the very structure of these sciences themselves.

In a previous lecture [8] we opposed to the purely materialistic, positivistic or quantitative conception of science, the search for the typical, the attempt intuitively to seize the original secret, the proper treasure of sensible reality. Even though this search be conducted by purely empirical means and with the help of concepts resolving themselves into the observable, it evidently implies a tendency to philosophy, to ontology; it implies an ontological orientation. But at the same time, science thus oriented by the philosophy of nature must keep itself from rising to the properly philosophical plane, since it must be held within the limits of empiriological analysis and of the empiriological vocabulary; that is to say, it must keep on resolving its concepts in the observable and not in intelligible being as such.

Nevertheless the need for these philosophical truths as *directive principles* allows us to speak here of subalternation improperly so-called, or of organic and vital subordination.

16. If we want to make a diagram of what has just been said, we may represent the situation in the following way:

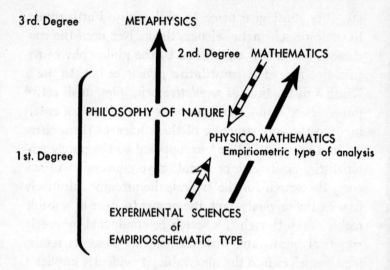

To the third degree of ideative visualization corresponds metaphysics. Within the generic unity of the first degree are two specifically distinct spheres: the philosophy of nature and experimental sciences of the empirioschematic type. These sciences are subordinated (improperly subalternated) to the philosophy of nature by reason of their principles (without forming a *scientia media*) and *regulatively*, not constitutively. They belong to the same generic degree but are specifically distinct from the philosophy of nature.

At the second degree of abstractive visualization, which is on another vertical line, the physico-mathematical experimental sciences (empiriometric type of analysis) are subalternated to mathematics, but here the subalternation is proper; not only by reason of principles but by reason of principles and object; with the result that these sciences constitute with mathematics an intermediary science: materially physical and formally mathematical. These sciences are astride two generically different degrees of abstraction: they belong to the first, the physical, degree of abstraction because they are materially physical, and they belong to the second degree (generically different from the first) because they are formally mathematical.

The intellectual or spiritual direction of the ancients is indicated on the diagram by an arrow pointing toward metaphysics and showing that, for them, metaphysical intellection was supremely regulative of knowledge and that all knowledge ultimately underwent this metaphysical attraction; the experimental sciences doing so via the philosophy of nature. For the ancients then, supreme regulation of all knowledge was exercised by metaphysics. To symbolize the spiritual direction of the moderns, on the contrary, we must draw another arrow showing that for them the supreme regulation of all knowledge is to be found on the side of mathematics.

If this diagram is correct, you can see that the empirioschematic experimental sciences are a locus of conflict between the tendency to subordinate knowledge finally to philosophy, and the other tendency to subordinate it to mathematics.

In winning their autonomy the experimental sciences of the empirioschematic order, which may also be called typological sciences, escape in some measure from the imperialism of mathematics, but they are subordinated in the same measure to the philosophy of nature,—not constitutively but regulatively. Philosophical truths give the scientist orientation and direction of the greatest importance without entering, for all that, into the notional or conceptual structure of science.

Applications to Biology

17. Applying the notions we have been discussing to the knowledge of the living organism, we can distinguish three types of biological knowledge; the distinction can be made theoretically at least, for in reality of course, these types are always more or less mixed with one another. First of all there will be an empiriometric or physico-mathematical biology, a biology which will tend ultimately to offer a mathematical interpretation of sensible data. We are still very far from such a science, but its first outlines may already be seen. Insofar as the scientist arrives at physico-chemical explanations of the living real he approaches this physico-mathematical biology, because the physico-chemical sciences themselves are a part of physico-mathematical knowledge and tend implicitly to be resolved as far as possible into mathematical concepts. Actually, non-living matter is the chosen and preferred field of empiriometrical explanations, but there is no reason why these explanations should not be developed in biology and even advance indefinitely therein. They bear on what may be called the material conditioning of life, the physico-chemical *means*

of life. If it is true that life uses physico-chemical instruments, equipment and means, then an empiriometric science of the living being concerned with these physico-chemical means themselves should be able to make endless and unlimited progress without exhausting for that the reality of the realm of life.

However, I think that this empiriometric knowledge must always remain a subordinate part, a *means*, an *instrument* of typological biology, so that the physico-mathematical explanation in biology will never lead to pseudo-ontology, to the closed world of mathematicism with its pretensions of giving a total explanation and reconstruction of the real. Even though this bio-mathematical discipline imply a tendency to mathematicism or mechanism, this tendency will remain inefficacious, precisely because this part of biology could never constitute an autonomous whole. To try to set it up as an autonomous whole would be to succumb to the illusion of a *living-being-less biology* (just as a certain experimental psychology pretends to be a "soul-less psychology" and a certain school of medicine, governed exclusively by laboratory reactions, "patient-less medicine").

Evidently if it be a question, not of the empiriometrical analysis we are speaking of here, but of the construction of an explanatory pseudo-ontology, mechanism is a temptation to which it does biology no good to cede. For example, a few years ago the cellular theory was interpreted in a materialistic sense: the organism was considered as a simple association of cells without its own substantial unity. This theory, very much in vogue for a while, has been discredited, re-absorbed by the natural play of scientific progress.

18. Above physico-mathematical biology which bears not precisely on life itself but on the material means, the physico-chemical means of life, above this biology I say, there would be a biology that may be called typological or *formally experimental biology*. This would have life itself for its object and bear on the living being itself; but for its analysis of life it would use empiriological and not ontological ways of thinking, notions and definitions. In a word it would resolve its concepts into the observable. This typological biology or formally experimental biology would have its separate, directive principles in philosophy; it would lean, so to speak, on implied philosophical concepts, but would have its own autonomous conceptual vocabulary, specifically distinct from that of philosophy because, to repeat, it would resolve its notions and concepts into the observable as such and not into intelligible being as such.

A science can be directed from without by another science, so it is admissible that the philosophy of nature perform a regulative function with respect to biology or any similar discipline without encroaching on its proper domain and in leaving it all its freedom and autonomy. For example, notions such as finality, vegetative activity, (*potentia vegetativa*), soul, (or, I would say, substantial tensor, *forma substantialis*), have an explanatory value in the ontological knowledge proper to the philosopher of nature. They are philosophical notions; thanks to them the philosopher of nature interprets experience and makes it intelligible.

Now a mind that thus possesses a certain philosophy of living nature, a certain philosophy of the organism, is oriented in a certain well-determined manner in its experimental research in histology or any other branch of biology. But, to the measure in which experimental science most perfectly realizes its own nature, notions such as finality, substantial tensor, *potentia vegetativa*, must not penetrate the formal structure of the scientific discipline; the experimental scientist must not invoke them as principles of explanation. The philosopher, yes, but not the experimental scientist: he may be directed, oriented by them; he must not invoke them as principles of scientific explanation. That seems especially important to me with respect to the concept of finality. We may ask what is the rôle of finality in biology; indeed, the question is always coming up; but I think we had best distinguish strictly, here, between *formally experimental biology* and *philosophical biology*, the philosophy of the organism. Having made this distinction, we can see that finality has a value or properly explanatory significance for the philosophy of nature, for philosophical biology, but has no properly explanatory value for formally experimental biology. And nevertheless it is present; we cannot deny that it is there but, in my opinion, we must say that it is present as an irrational element or as a *pre-explanatory condition* which the scientist is compelled to recognize, whose existence he must admit, but which does not enter into the structure of his explanation.

19. The third type of biological knowledge is this *philosophical biology* we have just spoken of, which is a particular chapter of the philosophy of nature and in which concepts will have their full intelligible freight, their whole speculative value without any forcing back toward the senses. In it explanation will be sought in terms of a *raison d'être* or principle of intelligibility and of essence; these explanations, as we have already explained, will not be able to enter into the detail of phenomena but will deal with the most general and fundamental realities presented by living being.

SECTION 2 · DEFINITION OF THE PHILOSOPHY OF NATURE

The Philosophy of Nature and Metaphysics

20. We shall now proceed to find out what the philosophy of nature is, and pass from the question *an est* to the question *quid est*. What is the philosophy of nature? How shall we define it?

This question has been treated in a most interesting manner, from the point of view of the ancients, by Cajetan in his opusculum *De subjecto naturalis philosophiae*. In this work Cajetan has clearly shown that the philosophy of nature is not a part of metaphysics nor properly subalternated to metaphysics. He has shown that the proper "subject" or object of the philosophy of nature is being as mutable, *ens mobile*, being taken under the formal reason or from the proper perspective of motion or mutability,—a proper perspective which restricts the notion of being without depriving it of its transcendental and analogical character. If you were to define the philosophy of nature by saying that its specifying object is an object of thought of the generic order, such as *corpus naturale*, bodies and their properties, you would not bring out the fact that it bears on being with the analogicity that being connotes. Its object is always 'being,' which is an essentially analogous object of thought permeating all generic and specific diversifications, but 'being' restricted by the note 'mutable' or 'moving,'—being insofar as it is mutable. That is why we are confronted here with a philosophy. Cajetan also shows that the expression *"ens sensibile"* which might be used and which is not illegitimate in itself, is nevertheless less formal and less philosophical than the expression *"ens mobile."* The latter expression frees us straightway from the snares of Parmenides and Melissus for, in asserting that mutable or moving being is the proper object of the philosophy of nature, we thereby affirm what Parmenides and his school denied: that being seen from the perspective of mutability is knowable and can be an object of knowledge.

Thus for the ancients, the philosophy of nature is indeed a philosophy since it is concerned with being, but it is not a metaphysics. This is the theme we stated and have repeated from the very beginning of these lectures. It is not first philosophy because it does not bear on being as being, being in its own intelligible mystery. The philosophy of nature is inferior to metaphysics; it is at the first degree of ideative visualization; it studies being as mutable, being under the conditions of poverty and division which affect it in that universe which is the material universe, being viewed from the outlook of the mystery peculiar to becoming, to mutability, to the movement in space where bodies interact, the movement of generation and substantial corruption which is the deepest mark of their ontological structure, the movement of vegetative growth wherein is made manifest the ascension from matter to life.

From this it is evident that for Cajetan, as for all Thomists, it is a serious error to confound the philosophy of nature with metaphysics. Do we need to repeat what we have already said on this subject? Metaphysics does not need to be complemented by the experimental sciences of nature for it does not bear upon mutable being but upon being as being. And the definitions used by the philosophy of nature intrinsically imply in their very intelligibility a reference to this or that determined sensory act; but this is not true of the notions and definitions of metaphysics.

Although the philosophy of nature is essentially distinct from metaphysics because of the basic characteristics of its generic type, yet it has a fundamental importance for metaphysics. For us humans,—(in angelic knowledge there could be no question of the degrees of abstraction),—the philosophy of nature constitutes the first germinal differentiation around which spring up all the other parts of philosophy. We draw our most abstract ideas from experience, consequently the philosopher first deals with the realm of being as moving, the realm of the philosophy of nature. If he proceeds in a properly human way, he will,—with regard to the inner structure of knowledge if not to his own procedure in time,—treat first of the philosophy of nature, taken at least in its major essential determinations, before going on to metaphysics. And therefore were we to suppress the philosophy of nature, exile it from the sphere of knowledge, as we have seen that modern thought does, there would be no metaphysics open on to things and on to the immensity of being; there would be no speculative metaphysics. There would be only a reflexive metaphysics,—reflexive and openly idealist like Brunschvicg's, which seeks spirituality in an awareness of scientific progress wherein the mind endlessly surpasses itself,—or reflexive and covertly idealist as is Husserl's and many neo-realists, —or reflexive and ineffectively realist like Bergson's, seeking within physico-mathematics a metaphysical content which is unknown to this science and is revealed only by the intuition of pure change. Or reflexive and tragic, as are so many contemporary metaphysics in which, in Germany especially, the spirit endeavors to rediscover the sense of being and existence in the drama of moral experience or in the experience of anguish.

Suppress the philosophy of nature and you suppress metaphysics as speculative knowledge of the highest mysteries of being naturally accessible to our reason. There is an involution of causes here: *causae ad invicem sunt causae*. Metaphysics is necessary for the constitution of a sane philosophy of nature to which it is surordinate; but inversely, metaphysics itself is soundly constituted only by presupposing a philosophy of nature which it uses as its material basis. The very nature of our mind is involved in this. Having no immediate contact with the real except through our senses, knowledge of the pure intelligible, knowledge at the highest degree of natural spirituality, cannot reach the universe of immaterial realities if it does not first get a hold on the universe of material realities. And it cannot grasp this universe, hunt out its proper object in it, if it be held as impossible that the mind have knowledge of the intelligible mixed with or shadowed by the sensible, knowledge inferior in spirituality which first attains to the being of things as steeped in mutability and corruptibility,—knowledge which prefigures metaphysical truth in the shadows of this first degree of philosophical knowledge. Without a philosophy of nature which is surordinate to the natural sciences and subordinate to metaphysics and which preserves the contact between philosophical thought and the universe of the sciences, metaphysics has no contact with things and can only fall futilely back upon the knowing or willing mind itself. In the order of material and dispositive causality, the wisdom *secundum quid* of the philosophy of nature, taken in its first positions at least, is a condition for speculative wisdom pure and simple, a condition for metaphysics.

And, conversely, without a philosophy of nature to as it were transmit rulings from above to the world of the natural sciences, metaphysics can no longer exercise over the latter its function of *scientia rectrix*. I mean that it remains ineffective either to orient toward a knowledge of wisdom, everything in the sciences of phenomena which aspires without attainment to an intelligible grasp of the real as such, or to judge and delimit the meaning and scope of whatever is subject in these natural sciences to the final regulation of mathematical entities. The immense and powerful mass of scientific activities, the human mind's marvelous endeavor to conquer nature experimentally and mathematically, is left without any direction or light higher than that of empirical and quantitative law, and is wholly cut off from the order of wisdom. It advances historically and it captivates men, but it no longer knows aught of speculative and practical wisdom.

The Philosophy of Nature and the Sciences

21. Cajetan was fully justified, therefore, in defining the philosophy of nature by this formal object: mutable being insofar as it is mutable;—*the moving*, as we would say today in Bergsonian language. But several points still remain to be cleared up. We have already said that the ancients failed to distinguish, or distinguished very in-

sufficiently, between the philosophy of nature properly so-called and the natural sciences. Due to the progress of these sciences, we must now stress this distinction without straining it. What then shall we say on this subject?

First of all, we must remember that the philosophy of nature and the natural sciences are *at the same generic degree* of abstractive visualization and bear equally upon sensible and mutable being.

Secondly we must keep in mind that there is a specific difference between these two types of scientific knowledge; a specific difference springing from the difference in their mode of defining: the one making use of empiriological analysis, the other using ontological analysis of the sensible real.

Finally we must remember that we are not dealing here with two sciences that are simply parallel and never come into contact with one another. Rather, as we have already said, between these two specifically distinct types of knowledge there exists the same sort of relationship as exists between the soul and the body: a relationship of complementarity despite their specific distinction. This comparison to the soul and body is deficient in this that despite the difference of nature between soul and body they constitute a single *substance*, which is specifically one. Obviously the philosophy of nature and the sciences are not the elements of a single substantial whole, and we have just said that they are specifically distinct. But, from the point of view of the integrity of the reality to be known, this comparison holds, because the universe of the sensible real is integrally known only by the meeting and collaboration of the philosopher of nature and the scientist.

Formal Objects and Formal Perspectives

22. Now if we want to clarify these notions and make them more precise we shall have to resort to scholastic terminology at its dryest, to Cajetan's teaching about formal objects and formal viewpoints (*rationes formales*).

Quite independently of our present interest in the question of the philosophy of nature, it is an excellent thing to clarify our thoughts on this doctrine of "formal objects" and "formal viewpoints" or proper perspectives of knowledge, for this is a highly important doctrine on which depends the whole specification of *habitus* and of the sciences.

Cajetan expounds this doctrine in his commentary on the I*a Pars* of the *Summa Theologica* (q. 1, art. 3). In this article he is treating of the relations that exist between theology, a human science, and the intuitive science of the blessed of which theology is a sort of impression and participation and to which it is subalternated.

Cajetan explains that we must first consider here what he calls the *ratio formalis objecti ut res* or the *ratio formalis quae*. These formulae are rather difficult to translate and the vocabulary of the ancients could do with a bit of renovating. So, since we are here concerned with the way in which things invite or, if I may so speak, "appeal to" the mind to understand them, solicit understanding, present themselves intelligibly to the knowing mind, may I propose,—it is just a suggestion,—such an expression as INTELLIGIBILITY-APPEAL to render this notion which the ancients called *ratio formalis quae* or *ratio formalis objecti ut res*, the formal perspective of reality or the formal perspective of the object as a thing.[9] That, says Cajetan is the *ratio rei objectae quae primo terminat actum illius habitus, et ex qua fluunt passiones illius subjecti*, the formal aspect of the reality presented to the mind, which the mind seeks first and foremost or which first and foremost invites the act of this or that *habitus* and from which are derived the properties of this or that subject of knowledge, for example:

>   *entitas* in metaphysica,
>   *quantitas* in mathematica,
>   *mobilitas* in philosophia naturalis.

Such is the intelligibility-appeal, the *ratio formalis objecti ut res*, the aspect under which the thing presents itself to the knowing mind, the intelligible face which it shows to the mind and by reason of which a first cleavage or differentiation is produced in our intellectual activity, a first determination of our mind's glance toward things and of our intellectual stable dispositions (*habitus*).

We may say then that the formal object of metaphysics is *ens sub ratione entitatis*; the formal object of mathematics, *ens sub ratione quantitatis*; the formal ob-

ject of the philosophy of nature, *ens sub ratione mobilitatis*. So we have here: 1. *ens* which may be called the
material object of knowledge; 2. *entitas, quantitas, mobilitas*, the intelligibility-appeal made by the thing, the
ratio formalis objecti ut res, the aspect or rather the
inspect, the perspective of intelligibility which the thing
presents primarily to our understanding; and 3. the two
taken together: *ens sub ratione entitatis, ens sub ratione
quantitatis, ens sub ratione mobilitatis*, which we shall
call the *objectum* or *subjectum formale quod*, the formal
object *quod*, which is the material object taken from this
or that formal perspective. Since we are calling the *ratio
formalis objecti ut res* "intelligibility-appeal," we shall call
the formal object thus determined "THE SPHERE OF
FUNDAMENTAL INTELLIGIBILITY."

To clarify this further, let us take the science of medicine for example; the material object there is the human
body but it is considered *sub ratione sanationis*, that is,
as capable of being healed. This *ratio sanationis* is the
ratio formalis objecti ut res or the intelligibility-appeal
of the thing. The whole, the human body taken according to this intelligibility-appeal, this perspective of reality,
is the formal object *quod* or the sphere of fundamental
intelligibility of the art or practical science of medicine.

23. But we cannot stop here. Besides these perspectives there is,—and here the thing starts to get interesting,—what Cajetan calls the *ratio formalis objecti ut
objectum*, the formal perspective of the object *as object*,
or the *ratio formalis sub qua*, the formal perspective

under which the object, otherwise determined by the *ratio formalis quae*, is attained by the mind. And we may translate this as "OBJECTIVE LIGHT." [10]

How does Cajetan describe this perspective? He says that this formal perspective is constituted by a certain type of immateriality, *immaterialitas talis*, a certain type or degree of abstractive immateriality, or again *talis modus abstrahendi et definiendi*, a certain mode of abstracting or defining. The objective light (*ratio formalis sub qua*) is the formal perspective of conceptualization. For example, *sine omni materia* for metaphysics; *cum materia intelligibili tantum* for mathematics, and for natural philosophy, *cum materia sensibili, non tamen hac*: abstraction from individual matter but not from sensible matter.

In the case of medicine, which we took as an example above, we would have to say that the objective light or the formal perspective *sub qua* is the abstractive immateriality peculiar to the first order of abstraction, proceeding according to the compositive mode proper to practical sciences.

24. A two-fold remark must now be made: 1°. The *objective light*, (that is, the type of immateriality or intelligibility according to which the knowing mind constitutes and conceptualizes its object, the perspective of conceptualization) plays a more formally specifying role than does the intelligibility-appeal (the perspective of reality) which otherwise determines the object and which is revealed by this typical illumination. For it is by the

object, insofar as it measures the act, that *habitus* are specified. The objective light,—the formal perspective *sub qua*,—has a more *formative* function than has the intelligibility-appeal (the formal perspective *quae*)—with respect to the object as such and at the same time with respect to the cognitive act. Thus the typical light, the conceptual perspective whose objectivity is purely intelligible or perfectly immaterialized (*modus abstrahendi et definiendi sine omni materia*) specifies metaphysical knowledge more exactly, more formally and more decisively than does the perspective of reality *"entitas"* or the formal doubling back on being itself according to which the real is considered in metaphysics.

2. It may happen that, given a certain sphere of fundamental intelligibility determined by the intelligibility-appeal of the thing, the corresponding objective light be diversified into several different objective lights each specifying a type of knowledge. In such a case it is clear that what ultimately specifies a scientific *habitus* is the formal perspective *sub qua,* the objective light, more than the formal perspective *quae*.

Such is the case for theology,—and this is Cajetan's point: theology has the same intelligibility-appeal, the same formal perspective of reality (*as does the beatific vision: Deitas ut sic*) and consequently belongs to the same sphere of fundamental intelligibility. The intelligibility-appeal, the *ratio formalis quae* of theology is deity as such, the deep depths of the divine nature; its sphere of fundamental intelligibility is *Deitas sub ratione Dei-*

tatis, God taken not according to the intelligibility-appeal of first cause, but according to that of deity itself. And yet the formal perspective *sub qua*, the objective light of theology is not the light of the beatific vision and of the science of the blessed; our theology proceeds from a special objective light: the light of divine revelation, not *as* evident as it is in glory and not *as* inevident, but simply *as* revealing: for the principles of theology are received from the intuitive science of the blessed by means of faith. In this case the intelligibility-appeal, the formal perspective of reality, has only a generic and not a specific value of determination, and the objective light corresponding to this intelligibility-appeal, (the formal perspective *sub qua* which corresponds to this formal perspective *quae*) also has a generic unity which is diversified into several species. The *lumen divinum* is divided first into *lumen divinum evidens*, which is the perspective *sub qua*, the objective light of the theology of the blessed; secondly into *lumen divinum revelans abstrahendo ab evidentia aut inevidentia*, the divine revealing light considered neither as evident nor inevident, which is the objective light of our theology; and finally *lumen divinum inevidens*, the non-evident divine revealing light which is the objective light of faith. Three different objective lights for one same sphere of fundamental intelligibility, for one same object intelligibly determined by the formal perspective of the object as a thing (*Deitas*).

Let's take some less lofty examples, much humbler

and merely approximative, but helpful to the imagination. Suppose that we have as material object a colored canvas, an *artefactum*. This colored canvas, this work of art presents itself to the mind with a certain intelligibility-appeal, say with the differentiating characteristic of a thing *painted for the sake of beauty*. (The canvas in question is a masterpiece.) The material object is a "colored canvas" (a concept which could apply just as well to a piece of oilcloth as to a painting); the intelligibility-appeal or the formal perspective of reality is the characteristic "painted for the sake of beauty"; the formal object thus determined is the painting. Or again, a man (material object) comes to my house, presents himself to me as such and such, as a friend, a tradesman, a creditor. Friend, tradesman, creditor, this is the intelligibility-appeal, the formal perspective of reality. Now, I would prefer to suppose that he is a friend of mine.

There is yet another thing to be considered in this man, namely the appropriate way of talking to him. He might be a sensitive or reserved friend with whom one's conversation should be guarded and subtle, or a familiar friend with whom one could converse casually; or he might be an unfortunate friend needing someone to weep with him, according to the precept of St. Paul, or a fortunate friend expecting to be congratulated. In each instance you have one same generically determined formal object: a man taken according to the intelligibility-appeal of friendship, but nevertheless you have different modes of talking to him. The notion of objective light corresponds to these different modes of conversation; it is the formal perspective *sub qua*, the way of dealing with the object, the way of entering into conversation with it. Likewise in the example of the painting: this painting must be looked at under a certain objective light. If it is an anatomical or botanical drawing we must consider its fidelity to nature; if it is a painting by Rembrandt we must take a completely different point of view: the point of view of the soul and predestination for which this painter is nostalgic and whose mystery he makes palpable. If it is a Picasso we should look at it from the point of view of the abstract reconstruction of objects. So there are very different ways of conversing with one same work formally determined as *res*, different objective lights by which to attain and understand that work. The necessity of this formal point of view, this objective light, is frequently overlooked; it is thought sufficient simply to look at a painting in order to judge it as such; whereas, to get to the heart of a work of art one must share the point of view of its maker in a certain way, and this point of view corresponds to the objective light. You have to accept a sort of postulate; you have, as it were, to trust the artist. Before you can judge a work of art you must take a certain point of view, adopt a certain intentional perspective. As soon as the artist has, in a word or two, explained what he was trying to do, *your judgement is oriented* and you are capable of judging the work of art in question. But if you make no such act of sympathetic acceptance, then you will never be able fairly to judge that work of art. We have need of an objective light which is more determining, more specializing than is the intelligibility-appeal of the object as thing, the *ratio formalis objecti ut res*, if our judgement is to be equitable.

25. The appropriate moment for such a summary having arrived, we can now sum up all the foregoing briefly by saying that normally, because of the normal correspondence between the reality to be known and the manner of knowing and conceptualizing, every intelligibility-appeal (formal perspective of reality) has a corresponding objective light (formal perspective of conceptualization) and vice versa. But the objective light is the more specifying and this correspondence is established in different ways.

Now, since it is the typical way in which the real presents itself to and solicits the understanding, the intelligibility-appeal of a thing can have a specific value by itself (either immediately, as the formal perspective *entitas* is a specifically metaphysical intelligibility-appeal, or consequent on the division of a generic appeal: as the formal perspective *quantitas* is divided into *quantitas continua*, specific intelligibility-appeal for geometry, and *quantitas concreta*, specific intelligibility-appeal for arithmetic). In this case the specifying objective-light corresponds to an already specified intelligibility-appeal. And it is according to this intelligibility-appeal (which we shall say is of *primary determination*) that the real is presented to the specifying objective light.

But the intelligibility-appeal of a thing, by itself, can also have a merely generic value (this is true of the formal perspective *deitas* and, insofar as they usefully illustrate our point, of the formal perspectives *friendship* and painted *for the sake of beauty* in the examples we used above). In this case it is the objective light alone that causes the specific determination of the object as object, without meeting in the thing any previous circumscription of specifically different spheres of reality. The specifying objective lights diversify the generic value of the thing's intelligibility-appeal and thus *select their own corresponding* intelligibility-appeals of specific value. It is according to the latter intelligibility-appeals (which may be said to be *induced* or of *second determination*) that the real is presented to these objective lights.

When the intelligibility-appeal of the thing is infinitely transcendent or infinitely simple, as is the formal perspective of deity, these intelligibility-appeals of second determination are simply repercussions or ideal reflections of the mode of knowing upon the object: we say that the specifying formal object *quod* (the intelligible sphere of second determination) of the vision and science of the blessed, of theology and of faith, is God according to the formal perspective of deity: in the first instance as *seen* and *known* with evidence, in the second as *known* be it with evidence or inevidence, in the third as *believed* without evidence. But to say this is simply to state the typical mode of knowing the object due to the specifying objective light.

In other cases,—in the case of the examples we were just using,—the intelligibility-appeals of second determination doubtless presuppose certain diversities of aspect in the thing itself and discover in it varied internal perspectives; but the latter are differentiated only upon the solicitation of the objective light and, as it were, in answer to different points of view on the same sphere of fundamental intelligibility. I am called upon by the friend who comes to my house or by a painting which is shown to me, to enter into a certain conversation and it is only as a dependent variable of the objective light illuminating the real and soliciting it to reveal itself that I may class in a particular typical line the aspects of their being which the friend or the painting thus reveals to me.

The Philosophy of Nature and the Empirioschematic Sciences

26. We can now apply what we have just said to the problem of the relations between the philosophy of nature and the natural sciences, more precisely *between the philosophy of nature and the non-mathematicized sciences of nature,* those which we have classified as empirioschematic. What shall we say about them now that we have these keys and understand these scholastic concepts? We shall say that the philosophy of nature and the sciences in question answer to the same intelligibility-appeal in the thing and consequently have the same formal object *quod,* the same sphere of fundamental intelligibility: this sphere of fundamental intelligibility is

mutable or moving being, being as mutable, *ens sub ratione mobilitatis. Ens,* material object; *mobilitas,* formal perspective of reality, the intelligibility-appeal of the thing; *ens sub ratione mobilitatis,* sphere of fundamental intelligibility. Therefore they have the same subject or object, the same sphere of fundamental intelligibility; that is why the philosophy of nature and these natural sciences are mutually complementary.

But this sphere of fundamental intelligibility, determined by the formal perspective of reality, has only a generic unity. And the philosophy of nature and the natural sciences are specifically different. What then is the source of this difference? The difference comes from the objective light, the formal perspective of conceptualization, the way of conversing with the object. Compared to the philosopher's, the objective light of the non-mathematicized natural sciences is dimmer, the chiaroscuro of empiriological conceptualization. And this minimally luminous chiaroscuro in the mode of conceiving and analyzing is the only light by which the detail of phenomena, of actions and reactions in sensible nature, can be attained. In a more intense light these details vanish, they are no longer visible for the light consumes them.

Therefore in both philosophy and science, the mode of abstracting and defining leaves aside singular matter but not sensible matter. But in one case this mode of abstracting and defining tends toward sensible being *as intelligible;* this is the objective light proper to the philosophy of nature; in the other it tends to sensible being precisely insofar *as* it is *observable*; this is the objective light proper to the sciences of phenomena.

If we want to find a Latin expression to designate the objective light which is proper to, and specifies the non-mathematicized sciences of nature, we may use *modus definiendi per operationem sensus.*

To this objective light corresponds an *induced* intelligibility-appeal (an intelligibility-appeal of second determination) which is precisely phenomenality. Phenomena are not special things; a phenomenon is not a certain thing or formal object of first determination, a certain stratum of knowable reality distinct from something else which is the thing in itself and constituting a world apart. Phenomena are simply the aspect in the formal object of primary determination, in the sphere of fundamental intelligibility proper to the first degree of abstractive visualization, which meets with a mode of defining and conceptualizing, an objective light that proceeds by resolution into sensory operation.

The specific formal object *quod,* the intelligible sphere of second determination thus characterized, must then be defined: *ens secundum quod mobile* (this is common to the philosophy of nature and the sciences) *sub ratione phenomenalitatis, id est sub modo definiendi per operationem sensus,* or again—our Latin gets to be a bit awkward here,—*ens secundum quod mobile sub lumine empiriologico*: mutable being considered from the point of view of the detail of observable phenomena or, in other words, mutable being seen under the objective light of the mode of defining by sensory operation.

The Philosophy of Nature and the Empiriometrical Sciences

27. So much for the non-mathematicized sciences; now for the mathematicized natural sciences, those which we have called empiriometrical. Here not only are the objective lights for the philosophy of nature and the sciences different: ontological conceptualization on the one hand, empiriological on the other, but furthermore the intelligibility-appeal issuing from the thing, the formal perspective of reality, differs too. For in the physico-mathematical sciences the intelligibility-appeal is *quantitas*, the same as it is for mathematics; in the philosophy of nature it is *mobilitas*.

Here we have sciences that *terminate* in the physical object, in the sensible real, but their formal object of primary determination, their sphere of fundamental intelligibility, differs from that of the philosophy of nature; it is at the same time determined materially in function of *mobilitas* and formally (or as to its proper degree of intelligibility) in function of *quantitas*. Such is the object of a *scientia media*, an intermediary science; and we may define this sphere of fundamental intelligibility as *ens mobile sub ratione quantitas*. These sciences have as their material object the object of physics, *ens mobile*, but they take it from the formal perspective of mathematics, *sub ratione quantitatis*. Now what about their formal perspective of conceptualization their objective light? It is the empiriometrical point of view, the mode of defining by sensibly effectuable means.

Thus the complete definition of the intelligible sphere of the empiriometrical sciences will be *the moving* (mutable being), considered from the point of view of the proper intelligibility of quantitative relations, or from the point of view of the detail of measurable phenomena; that is, by the objective light of a mode of defining and conceptualizing which is carried on by sensibly effectuable measurements.

Finally, if we want a formula including both empiriometric and empirioschematic spheres of empiriological knowledge, we may say that the natural sciences, be they physico-mathematical or purely experimental, have as their object moving being with the intelligibility-appeal of mutability itself, or with the intelligibility-appeal of quantity, but always from the point of view of the detail of phenomena, or as observable and measurable, and not as intelligible: *ens mobile secundum quod mobile aut secundum quod quantum, sub modo definiendi per operationem sensus.*

Proposed Definition of the Philosophy of Nature

28. Now it is easier for us to define the object of the philosophy of nature. Its intelligibility-appeal (*ratio formalis quae*) is the moving, or mutability; its objective light (*ratio formalis sub qua*) is an ontological mode of analysis and conceptualization, a way of abstracting and

defining which, the while it refers intrinsically to sensory perception, aims at the intelligible essence; in this it differs specifically from the natural sciences.

The object of the philosophy of nature in all sensible things is not the detail of phenomena but intelligible being itself as mutable, or again, the differences of being which it can detect,—in aiming at the intelligible nature but without eliminating reference to sense data,—in the world of ontological mutability.

The sphere of intelligibility proper to the philosophy of nature is therefore *ens secundum quod mobile, sub modo definiendi per intelligibilem quidditatem (et non per operationem sensus), seu sub lumine ontologico.*

In this section we have defined the philosophy of nature, we have tried to determine in a precise and technical fashion what this philosophical science is, as compared to the sciences that are concerned with phenomena.

The Philosophy of Nature and Facts

29. Now we shall take up an extremely important question of method which depends on the principles we have just set forth. This question is that of knowing on what kind of facts the philosophy of nature should rest. We have just distinguished in the same fundamental sphere of intelligibility, two different types of intelligibility, two different objective lights or conceptual perspectives: the objective light of the philosopher and the objective light of the scientist. Both of these belong to the first degree of abstractive visualization and rest upon

sensible facts; but they do not do so in the same way. That is the problem of method which we must now investigate.

A fact may be said to be a well ascertained existential truth. A truth is expressed in a judgement which links together two objective concepts. So a fact implies that a connection of objective concepts exists *a parte rei*.

This manner of speaking brings out the truth, that what we call a *fact* inevitably implies the activity of the mind,—judgement. Of course I do not mean to say that the mind distorts things in judging them; that often does happen; we frequently do extrapolate. Yet this is in no way necessary! What I mean is that a fact is not inscribed on the scientist as a graph is automatically recorded on a chart. A fact implies discernment, an act of the mind. It simultaneously implies a judgement made by the mind and, if the fact belongs to the first order of abstraction,—a perception, a sensory intuition. Take, for example, the most banal of facts: snow is white. To state this proposition is to deal with a sensible human experience in which the intellect is alert and involved. Given certain existential data the mind distinguishes the objective concept, snow, and the objective concept, white; at the same time as it distinguishes them by the 'first operation of the mind,' it identifies them in a judgement. We have here a judgement establishing a connection between two abstract concepts and made at the dictates of sense intuition.

This being true, it is clear that there will be as many

different degrees or orders of facts as there are orders of abstractive visualization for objective concepts. In other words, the very discernment of a fact takes place at a certain degree of abstraction. Facts are not all of equal rank, they are not all grouped at the same level in the market-place of sense experience so that the different sciences may come and pick out the wares they need. Facts themselves enter into the hierarchy of our knowledge. Wherefore there are common sense facts, scientific facts, by which I mean facts which are of interest to the natural sciences of phenomena, mathematical facts, logical facts, metaphysical facts, etc.

From these premises it follows that there are philosophical facts that are much more simple, much more general, much more evident and certain than the facts which are called scientific, i.e. the facts handled, linked and interpreted by the natural sciences. For, as science progresses, these latter facts, especially in the physico-mathematical sciences, become points of contact of the real with increasingly complex constructions which have previously been set up by reason.

In his *Théorie Physique*, Duhem remarks that common observation is more certain than are scientific experiments. This is a curious remark for Duhem to make: "An account of a physical experiment does not have the immediate, relatively easily controlled, certitude of vulgar, unscientific testimony; less certain than the latter, it nevertheless has the advantage over it in point of the number and precision of the details it makes known;

therein lies its true and essential superiority." [11] Now it cannot strictly be said that the philosophical facts we are talking about here are facts which result from common observation; they are primordial pre-scientific facts, (if science is taken to mean the sciences that are devoted to the analysis of phenomena), but the pre-scientific observation is criticized and judged in the light of philosophy, in the light of philosophical principles and knowledge. There is a philosophical criticism of facts just as there is scientific criticism of facts. (This criticism of facts, observations and experiments is, as you know, an integral part of scientific work.) And when a fact which is the result of absolutely general observation has been judged and criticized by philosophy, it can no longer be called a fact of common observation, for the light of philosophical judgement and criticism has intervened to make it a philosophical fact in the strict sense of the word. The fact that something exists, that a multiplicity of things exists, that knowledge and thought exist, that becoming exists, these are all philosophical facts.

So in the genus common to the philosophy of nature and the sciences, that is, in the genus of the first order of abstraction we must distinguish philosophical facts of a different level than scientific facts and corresponding to the type of conceptualization specific to the philosophy of nature; for example, *change and becoming exist* (this is a philosophical fact and belongs to the first order of abstraction), *the continuous exists, successive duration exists.* Or again, the fact that *one substance changes into another*, is a fact to which primordial observation, judged and criticized by the philosopher, attests before any properly scientific elucidation; in the case of nutrition for example, where we see food become our own flesh, or in death where we see a living organism become something inanimate; it is not necessary to have studied much biology to know that much. The philosophers of the stone age were able to observe that there is an essential difference between a living and a non-living being and, therefore, that there is in such cases a substantial change. That one substance changes into another is a philosophical fact.

On the other hand there are, at this same degree of abstractive visualization, in the same genus of the first order of abstraction, scientific facts such as those that you may see gathered together in treatises on physics, chemistry, biology, etc.

30. Let us continue our investigation. Obviously since the means must correspond to the end,—(it is a fundamental axiom that the order of means corresponds to the order of ends)—philosophical knowledge of sensible nature must make use of nothing but facts of the same order. The philosophy of nature must make use of philosophical facts, that is to say, facts established and judged by the proper light of philosophy, because the more does not come out of the less; a fact can give only what it contains and philosophical conclusions can only be drawn from premises or facts which themselves possess philosophical value. Yes, but what will be the relation between

the philosophy of nature and scientific facts? Here we must point out two opposite errors into which it is unfortunately very easy to fall; this is one of the points where the greatest vigilance must be exerted in the elaboration of philosophical knowledge.

A first error consists in asking philosophical criteria of *brute* scientific facts. By a brute scientific fact I mean a scientific fact which has not been *treated* philosophically. Scientific facts in their brute state are not by themselves of interest to the philosophical type of explanation; as long as they are illumined only by the light which first made them discernible in the real and utilizable by the scientist, these facts are of interest only to the scientist, not to the philosopher. The scientist has the right to appropriate them for himself,—private property, no trespassing,—to claim them as his own; to say: no, these are mine, they are not yours; there is nothing in them for you but they are valuable to me for my scientific conclusions; you have no right to draw philosophical conclusions from them. It is an illusion to think that, by appealing to scientific facts on which no philosophical light has been cast, we can put an end to any philosophical dispute. That, it seems to me, is Father Descoqs' error in his book on hylomorphism.[12] He has, with very praiseworthy erudition, collected a great number of scientific facts, but from these facts as such he has tried to draw philosophical conclusions. Brute scientific facts tell us nothing about the question of matter and form; left in their brute state our only honest conclusion is to say that we know nothing about this question since they tell us nothing. It is not surprising that Father Descoqs' inquiry should have had such disappointing results.

The second error consists in rejecting scientific facts and trying to construct a philosophy of nature independent of them and isolated from the sciences. Now note that this is an inevitable tendency if the philosophy of nature be confounded with metaphysics; for in this case the philosophy of nature claims for itself the same freedom with respect to the detail of scientific facts as is possessed by metaphysics. In this eventuality we would have no metaphysics of the sensible but we would run the risk of having a metaphysics of ignorance. We would have a wisdom in a rudimentary, puerile state, (such cases of unbalanced growth are frequently found in the human mind in our day), but wisdom nevertheless, for a child can be in the right as against a philosopher; for example, a child who holds to the principle of causality as against a philosopher who denies it. We would have a rudimentary, puerile wisdom confronting an adult science, armed head to foot. Naturally such a wisdom would find itself in the inferior position.

31. In dealing with the problem of the relation of the philosophy of nature to scientific facts there are, then, two errors to be avoided. How can this problem be solved? To me the solution seems obvious: that the philosopher use scientific facts only on the condition that he *treat* them philosophically, deliver them of the

philosophical values with which they are pregnant, draw from them facts that have philosophical value. Philosophical facts which are the proper matter of the philosophy of nature may have two sources, both of which must be philosophically interpreted: 1° human, primordial, pre-scientific experience or 2° science, the immense domain of scientific observations and facts by means of which philosophical truths as yet unknown may be discovered or previously established philosophical facts may be confirmed, (e.g. the fact that nature gives us examples of substantial changes is confirmed by the analysis of truths established by chemistry and physical chemistry, by the physics of radio-activity, experimental biology, etc., provided that the scientific facts in question be philosophically judged and interpreted). In a word, philosophy may convert into its own substance matter that is foreign to it.

By relating scientific facts to philosophical knowledge previously acquired elsewhere and to the first principles of philosophy, by submitting them to the objective light of philosophy, we can draw from them an intelligible content that can be handled by philosophy. We can discern what ontological value these scientific facts possess, disengage philosophical facts from them by an original abstractive operation and by the activity of the *intellectus agens*. The problem of the relation of the philosophy of nature to scientific facts is not to be solved by using these facts in their brute, untreated state, nor by suppressing or neglecting them, and even less by forcing them, but by drawing philosophical facts from the gangue of sci-

entific facts,—as the *intellectus agens* draws intelligible
objects from sense experience.

One of the difficulties entailed by such a philosophical treatment is that frequently, especially in the physico-
mathematical sciences and in the very highly mathematicized branches of these sciences, it is very hard to distinguish between scientific fact and scientific theory. It is
all very well to say that in principle the outlook or perspective of fact is clearly distinct from that of theory. In
the first perspective the intervention of the intellect with
its most delicately refined constructs is always directed
to the discerning and formulating of existential positions
furnished by sense intuition (facts of the physical order)
or conceived by analogy with what is furnished by sense
intuition (facts of the mathematical order, logical order,
etc.). But in the perspective of theory, intellectual activity is directed to the discovery of causes, laws, explanatory
reasons.

Now actually, in the concrete movement of scientific
work, these two orders are constantly inter-mixed: there
is a continual circulation from fact to new theories which
it serves to construct, and from theory to new facts which
it serves to discern. It has often been remarked that the
facts which are immediately observed by the scientists
themselves presuppose a certain number of already ad-
mitted theoretical propositions relating first of all to
sense intuition and then to the thing to be measured,
the means of measuring, the instruments of measurement. As to the other mediately established scientific

facts, they result either from the coincidence of verified data with a previously established theory, or from an explanation which seems to be the only possible one. The interpretation of facts and theories is at its most complex in the physico-mathematical sciences. Sometime the mathematical elements may amount only (at least schematically) to grasping the physical and then we have what may properly be called a fact, *a fact with a real reference*, usable as real. Such, for example, is the *existence* of a material microstructure and of elementary particles which may be called whatever we like: atoms, electrons, etc.

Sometimes, at the other extreme, the physical enters only as a simple discriminatory element with respect to the theoretical constructions whose proper value is in their mathematical amplitude and coherence. Here the physical is simply a foundation for entities which have been reconstructed for the sake of mathematical explanation; such for example is the case of the *nature* attributed to electrons, atoms, etc., be it Bohr's electron or Schrodinger's, or to the waves of wave-mechanics. These are beings of reason whose foundation is real, and which *hide* reality even while they make it known. Such "facts" are improperly so-called; they are *facts with a symbolic reference* which the philosopher may find useful to create a mythical or symbolical representation to provisorily imagine things in a certain way in order to bring his philosophical conclusions into harmony with scientific imagery. In this part of the philosophical work

we are not dealing with knowledge properly so-called but
with a zone of essentially provisory and changing opinion.

For all these reasons, not only because of the zone
of physico-mathematical myths, but also and especially
because of the perpetual renewal of scientific ideas and
scientific language, because of the unceasing discovery
of new facts properly so-called, of new facts with a real
reference, we are compelled to conclude that since the
philosophy of nature needs the sciences for its completion and needs to draw confirmatory or illuminating
philosophical facts from the material of scientific facts,
it must therefore submit to a certain law of aging and
renewal. Not of course of substantial mutation! There is
a substantial continuity between the philosophy of nature as Aristotle saw it and as we see it; but it has undergone many changes on the way; it has grown old many
times and been many times renewed; as knowledge it is
much more dependent upon time than is metaphysics.

And this is indicative of the difference in their formal
objects, their formal values. We may say, if you like,
that a metaphysical treatise can come down through the
centuries and, if it be pure, defy time. Actually such a
treatise always contains allusions to the state of the
sciences in its day, to human opinions, etc.; but if it were
pure, it would defy time. Aristotle's *Metaphysics* will
never be out of date. But how long can a treatise in
experimental physics or biology live? Twenty years, ten
years, three years, two years, the lifetime of a horse, a

dog, a May-bug larva, a beet, a carrot, nay, even a day-fly. And a treatise on the philosophy of nature? Well, I think that a treatise on the philosophy of nature can, at the maximum, last a man's lifetime, fifty years, seventy years, *si autem in potentatibus, octaginta anni,*—and this provided that its successive editions, if it has them, be periodically brought up to date. For a treatise on the philosophy of nature must necessarily have intimate contact with the natural sciences, and these sciences are subject to much more rapid renewal than is philosophy.

The Contemporary Renaissance of the Philosophy of Nature

32. We have been speaking of the philosophy of nature considered as an abstract epistemological type. But we must add that today we are witnessing a sort of renaissance of the philosophy of nature. This renaissance parallels the retreat of the positivistic conception of science. Some biologists are expressly beginning to turn to philosophy for a deeper understanding of the living organism: sufficient to mention Hans Driesch's work on the philosophy of the organism,[13] Hans André's *"Urbild und Ursache in der Biologie,"* [14] and in France the *"Cahiers de Philosophie de la Nature"* founded by Dr. Rémy Collin.[15]

The magnificent contributions for which physics is indebted to Lorentz, Poincaré and Einstein on the one hand, and to Planck, Louis de Broglie, Bohr, Dirac and Heisenberg on the other, have also renewed and stimulated in this science the sense of the ontological mystery of the material world.

The major disputes and discoveries in modern mathematics concerning axiomatic method, the transfinite and the theory of number, the continuous and the transcendent geometries, are in need of philosophical clarification towards which the works of Russell, Whitehead or Brunschvicg constitute only a rather uncertain beginning. On the philosophical side the ideas of Bergson and Meyerson in France, of the phenomenologists in Germany (notably of Max Scheler) and of the Thomistic renaissance, have prepared the way for research dependent on an ontological knowledge of the sensible real. Whether or not this research will be directed toward a solidly founded philosophy of nature depends upon the activity of the Thomists.

Here we must beware of what we have elsewhere called dangerous alliances [16] and the temptation to an over-easy concordance, wherein the essential distinction between the empiriological and the ontological vocabularies is disregarded. This is an especial danger in regards to the relation of the philosophy of nature with the physico-mathematical sciences. For, as we have already observed, the latter in their most highly conceptualized theoretical branches reconstruct their universe by means of mathematical beings of reason founded in the real, by means of myths or symbols which as such have no connection with the real causes dealt with by philosophers.

33. But, so much having been said, we must also

take note of the very significant affinities which make modern science, despite the vast patches of shadow that still overlie it, more synergetic with Aristotelian-Thomistic philosophy of nature than is ancient and mediaeval science. We will not speak of the biological sciences, where the demonstration of this thesis would be too easy. The Cartesian concept of the world-machine and of matter identified with geometric extension; the Newtonian conception of an eternal framework of space and time independent of the world, of the infinity of the world, the pseudo-philosophical determinism of the Victorian physicists, all these dogmas have had their day. Contemporary science's ideas on mass and energy, the atom, mutations due to radio-activity, the periodical table of elements and the fundamental distinction between the family of elements and that of solutions and mixtures, dispose the mind, (I say dispose, for to say more than that all these materials would have to undergo properly philosophical treatment), dispose the mind to restore to their full value the Aristotelian notion of *nature* as the radical principle of activity, the notion of *substantial mutations* which is the foundation of the hylomorphic theory, and the notion of an *ascendant order* of material substances much richer and more significant than ancient physics ever surmised.

The idea of evolution is one which science itself does not handle without danger, and whose dissolvent power on the intellect Goethe denounced, but it would be futile to deny its definitive successes and its fertility. This idea, which sound philosophy can and must purify of its powers to delude, illustrates in a singularly striking way the fundamental notion, so often pointed out in these lectures, that the philosophy of nature is the philosophy of being in becoming and of the moving.

Looking upon our world wherein all is in motion, more so in the invisible atom than in the visible stars, and wherein motion is the universal mediator of interaction, the philosopher sees it to be wholly pervaded and, as it were, animated by the sort of participation of the spirit in matter which we call intentionality.

Its hierarchy has been reversed: the atomic world and not the celestial spheres is now the basis of time. The center of the physical world is no longer the sublunary globe surrounded by eternally rotating bodies that are both incorruptible and divine; rather is it the human soul, living its corporeal life on a tiny precarious planet, which is the immaterial and spiritual center of this physical world.

And this world is a world of contingence, of risk, adventure, irreversibility; it has a history and a direction in time. Bit by bit the giant stars grow smaller, are consumed and burn themselves out; for billions of years an enormous, original capital of dynamism and of energy has been tending toward equilibrium, using itself up, spending itself lavishly, bringing forth marvels in its rush toward death. The principle of entropy has been much abused by philosophers, but we nevertheless have the right to note this deep meaning which agrees so well with Aristotle's philosophical, not astronomical, notion of time: *quia tempus per se magis est causa corruptionis quam generationis.*[17] And we have also the right to point out how the natural exception which the least of living organisms makes to the law of the degradation of energy, (which applies, however, to the whole material universe) marks most significatively the threshold where something weightless, endowed with a singular metaphysical destiny and called the soul, empierces matter and opens up within it a new world.

In its way and with admirable precisions, science confirms that great idea by which the Thomistic philosophy of nature sees, in the universe of living and non-living bodies, an inspiration and an ascent from one ontological degree to another toward forms of increasingly complex unity and individuality, and of increasing interiority and communicability at the same time; an ascent towards that which is no longer just a part in this vast universe but is itself a whole, a stable universe open to others through intelligence and love:—the person which, as St. Thomas says, is the most perfect thing in all of nature.

In unriddling the picture of the mysterious universe given to it by the natural sciences, the philosophy of nature discerns at the heart of what may be called the tragedy of prime matter, a tremendous impulse to *answer*, instinctive at first, then stammered and then, in human beings, expressed in words; an impulse to answer to another Word which the philosophy of nature itself does not know. But Metaphysics knows it. By casting philosophical light upon the universe of the sciences, the philosophy of nature discerns therein an intelligibility which the sciences themselves cannot reveal to us. Disclosing, in sensible being known as mutable, analogical beginnings of the more profound truths and realities which are the proper object of Metaphysics, the philosophy of nature, precarious wisdom *secundum quid,* holds at the first degree of abstractive visualization, in the generic sphere of intelligibility nearest to the senses, the office of regulator and unifier of wisdom. Indispensable mediator, it brings into accord the world of the particular sciences, which is inferior to it, and the world of metaphysical wisdom which is above it. It is there at the very basis and outset of our human knowledge that the great law concerning the hierarchical and dynamic organization of knowledge, on which our intellectual unity depends, first comes into play; there, at the heart of the sensible and changing multiple.

IV

Maritain's Philosophy of the Sciences
By Yves R. Simon

The upholders of the Thomistic revival which began late in the nineteenth century were soon confronted with the following challenge: Because the philosophical principles of Thomism had been established at a time when positive science was in its infancy, it was asserted that Thomism was forbidden ever to deal successfully with the problems of our time. There could be no provision made in the system of St. Thomas for the interpretation of either the results or the spirit of modern science, both of which influence so deeply the very statement of our philosophical problems. The collapse of Aristotelian physics had entailed the general ruin of the Thomistic philosophy; against this verdict, rendered at the time of Galileo and Descartes, there could be no appeal. Thomism was at best a remarkable phase in the development of Western thought. If something of it could be revived, it was a certain inspiration, a certain aspiration, a certain frame of mind, but not any part of the systematic synthesis actually known under the name of Thomism.

Such was the only possible attitude for those who did not believe that any part of philosophy is independent of the data of positive science. Less radical-minded persons were willing to make an exception for metaphysics, considering that our knowledge of the one, the true, and the good is little affected by what happens in physics and mathematics. But when there is a question of cosmology, psychology, even of logic, the restoration of a philosophy conceived in the Middle Ages was deemed plainly impossible. The result was a number of eclectic constructions in which St. Thomas was permitted to supply a few general truths but not any refined and detailed achievement.

On the other hand, scholars convinced of the perennial truth of St. Thomas's philosophy were engaging in an ambiguous task: that of finding points of agreement between the teaching of St. Thomas and that of modern sciences. In the domain of psychology in particular, there is quite a literature about St. Thomas corroborated by the most modern and positive research.

As a matter of fact, in order to know how far Thomism was affected by modern developments in the positive sciences, a group of preliminary questions had to be investigated. What about the object of philosophy? Has philosophy any distinct object? What about the unity of philosophy? Is philosophy a science or not? One science or several? What is the significance of the distinction between philosophical and positive knowledge? Is it a necessary and everlastingly indispensable distinction,

as the science of the *ens mobile seu sensibile*. The physical object is both intelligible (*ens*) and observable (*mobile seu sensibile*). Neither of these opposite characteristics can be disregarded without its specific nature being destroyed. Leave out the words *mobile seu sensibile* and we are no longer dealing with something physical. Leave out the word *ens* and we fall below the level of intellectual knowledge.

Yet physical thinking, while bound to adhere to the two aspects of its object, can put a particular emphasis on either one. If the emphasis is put on *ens*, we have a form of knowledge both ontological and physical, a philosophical physics, a philosophy of nature. If the emphasis is put on *mobile seu sensibile*, we have a discipline of a physical and non-ontological character, an empiriological science. This point must be insisted upon: the privilege granted to either pole of the physical object is only a matter of emphasis. The philosopher of nature is not a metaphysician, and his definitions ought to imply some reference to data of sense experience. On the other hand the empiriologist is not a mere dealer in sense experiences, for the observable regularities with which he deals owe their constancy and their consistency to their being organized by some *ratio entis*. In this connection it is fitting to stress the felicitous character of this newly coined expression, *empiriological* sciences. Speaking of empirical sciences is objectionable, though customary, since empiricism is said in contradistinction to scientific knowledge. Empiriological sciences are not mere empiricism,

but a system of experience organized by an essential reference to a principle of intelligibility, ἐμπειρία μετὰ λόγου.

How physical thinking organizes itself around either
pole of its object can be best evidenced by investigating
the way physical definitions are constructed and justified.
A typology of physical concepts is the real key to the
opposition between philosophy of nature and positive
science.

Let us try a rigorous ascertainment of the meaning of
a word found both in philosophical and in positive contexts. The example chosen may be very simple. To the
question *what does the word man mean?* the answer will
be "rational animal"; now, none of the elements of this
definition presents a character of irreducible clarity. Take
one of them, for instance, animal. What does this word
mean? A correct definition would be: "a living body endowed with sense knowledge," and these are so many
terms which badly need clarification. Take one of them,
for instance, "living." I would say that a body is a living
one when it moves itself, when it is the active origin of
its own development. If we go any step farther, we go
beyond the limits of physical thought. In order to render
the idea of life clearer, we would have to define it as self-
actuation. The concept of self-actuation does not imply
any reference to the proper principles of corruptible and
observable things: it is a metaphysical concept. Its elements are identity and causality. Identity is the first property of being. Causality can be analysed into potency and
act. Identity, potency, and act are so many concepts

directly reducible to that of being, which is, in an absolute sense, the first and the most intelligible of all concepts. We have reached the ultimate term of the analysis,
the notion which neither needs to be nor can be defined
and which does not admit of any beyond.

This is the kind of analysis that the word *man* suggests when it is used in certain contexts. Everybody
would agree that a discourse which demands such an
analysis is a philosophical one. But the same word *man*
is often used in contexts which neither demand nor could
stand such an analysis. I happen to find the word *man* in
a treatise on zoology: explaining it in the way we did just
now would seem perfectly ridiculous. An analysis whose
term is the concept of being has obviously nothing to do
with the behavior, the method, the spirit and the principles of the whole discipline we call zoology. Should a
univocally-minded philosopher try to enlighten a zoologist by giving him explanations about self-actuation as
a particular form of relationship between potency and
act, no doubt the zoologist would burst into laughter and
declare that all these stories are perfectly nonsensical for
him as a scientist.

The zoologist would be right and the philosopher
would be univocally-minded. Both philosopher and
zoologist consider man, but they have a different way of
defining objects and of answering the question *what does
it mean?* For the zoologist, man is a mammal of the order
of Primates. How would he define such a term as mammal? A vertebrate characterized by the presence of special

glands secreting a liquid called milk. How is milk defined? In terms of color, taste, average density, biological function, chemical components, etc.

Here the ultimate and undefinable element is some sense datum; it is the object of an intuition for which no logical construction can be substituted and upon which all the logical constructions of the science of nature finally rest. In some cases, the explanation of a positive definition quickly demands recourse to sense experience. This often happens in the least elaborated parts of science. The elaboration of scientific concepts generally postpones the time when the recourse to sense intuition appears indispensable. But sooner or later it always imposes itself unmistakably. It is the possibility of being ascertained through sense experience which gives the concept its positive meaning. Every concept is meaningless for the positive scientist which cannot be, either directly or indirectly, explained in terms of sensations.

The philosophy of nature can be defined as a physical consideration whose conceptual instruments call for an ascending analysis, positive science as a physical consideration whose conceptual instruments call for a descending analysis. The very opposition of the two analyses provides an invaluable rule for the determination of the point of view prevailing in our studies about nature. Let us think of the ambiguous literature which stands on the borderline between philosophy and positive science. When a philosopher informed of positive science or a scientist interested in philosophy considers philosophical problems raised by the study of positive questions, the philosophical and the positive point of view appear successively in his expositions; generally the writer is not aware of the shift. The resulting confusion can easily be removed provided we carry out the analysis of a few key concepts. According as this analysis goes up or down, according as the concept demands to be explained in more and more characteristically ontological terms or in terms which refer more and more directly to definite experiences, we know whether we have to do with a philosophical or a positive treatment.

*　　*　　*　　*　　*

This description of positive science as a consideration of the *ens mobile seu sensibile* which puts the emphasis upon *mobile seu sensibile* and centers around the observable aspects of things throws a novel light on the notion of the science of phenomena. Let us have a glance at the adventurous history of this notion.

At the dawn of Greek philosophy, a science of phenomena is deemed impossible both by Parmenides and by Heraclitus. Science demands an unchangeable and necessary object; the phenomenal universe shows only a stream of changing appearances. The phenomenon, owing to its mutability, is thoroughly uncongenial to the spirit of scientific knowledge. This negation persists in Plato. The phenomenal world is the object of a merely opiniative knowledge; science finds its object in a transcendent world of numbers and ideas.[2] With Aristotle the picture is quite different. Aristotle realizes that there are immutable types immanent in the physical world: these are universal natures which reveal themselves through the regularities that are observed in the very order of phenomena. Accordingly, the phenomenon no longer has the character of an enemy of scientific thought. It is the phenomenon which, through its regularities, leads the scientific mind to its object: the universal types of things, their essences, their forms of being. The science so defined is a philosophy of nature, an ontology of the physical world. It does not reach its end until it is able to answer the question "What *is* the thing under consideration?" Neither Aristotle nor any of his Thomistic followers has ever construed the unwarranted idea of an intuitive perception of essences. Yet their scientific ideal is definitely attached to the disclosure, the understanding of the intelligible types immanent in the observable world. However essential may be the observation of phenomena in such a science, this science is by no means a science of phenomena. It is exclusively, or rather claims to be—for Aristotle did in fact perform great achievements in empiriological disciplines—a science of the essences located beyond the phenomena.

It can be safely said that the science of phenomena did not receive any epistemological charter before Kant. The charter it was given by Kant is an idealistic one. Hardly conscious of its nature in the era preceding the Kantian *Critique,* the science of phenomena, from then onward, was to be acknowledged as a distinct and fully legitimate epistemological species. But how is the old problem answered in the *Critique* of Kant? What sort of solution is given to the difficulty resulting from the sharp conflict between the requirements of the scientific spirit —necessity, universality, intersubjectivability—and the most obvious characteristics of the phenomenal world, its endless diversity, its thorough unsteadiness? There can be no doubt about it: the principles which, according to Kant, organize nature, do not lie in nature, but in the mind. The scientific object, with its characteristics of orderliness, determination, and universality, results from the application of mental categories to the diversity of sense-experience data.

Most men of science, ever since the Kantian reformation, have assented to the fundamentally idealistic view that the characteristics of the scientific object, its aptitude to fit in an intelligible system and, above all, to comply with the requirements of causal identification, are a proper effect of the constructive or synthetic activity of the mind. This stubborn adherence to an idealistic justification of positive science conflicts strikingly with the spontaneous realism of scientific thought. Men of science, willingly or not, receive their philosophical ideas from philosophers; they could not rid themselves of idealistic prejudices while philosophers were teaching idealism as the only doctrine that may account for the unquestionable ability of the mind to treat in an orderly and causal manner the universe of phenomena.

In his dealing with phenomena, Aristotle has no other

purpose than that of utilizing their regularities in order
to know essences. Maritain calls *dianoetical intellection*
the act of the mind which penetrates an essence and perceives what the thing is. For instance, the philosophical
definition of man as analysed above expresses an intellection which, inexhaustive and non-intuitive though it is,
has succeeded in penetrating the whatness of human
nature. We know that such a triumph of the theoretical
intellect is a rare achievement. In most cases we cannot
disclose the essences of sensible things in their specificity,
we cannot accomplish a dianoetical intellection of their
whatness. All we can do is to distinguish them through a
definition calling for a descending analysis. The intellection expressed by such a definition does not imply any
penetration of the physical essence, it only implies a circumscription of it within a steadily connected ensemble
of observable regularities. Nobody can say what the
essence of silver is; yet silver is a perfectly distinct chemical species. The undisclosed essence called silver is
clearly and certainly distinguished from any other
essence [3] by the system of observable regularities which
taken together belong exclusively to it. In this connection
let us call attention to a difficulty often experienced by
positive scientists when they try to give their definitions
a logically satisfactory form. We include in the definition
of silver the property of melting at 960.5° centigrade, the
property of boiling at 2000°, etc. But in the proposition,
silver melts at 960.5°, what does the subject, silver, refer
to, if not to something which is specified precisely by the

fact that it melts at 960.5°? The vice of circularity seems inevitable. The statement that silver melts at 960.5° resembles very much the statement that a black cat is black. Or, if we wish to avoid mentioning the predicate in the logical subject, we are confronted with a host of predicates hailing upon nothingness as a subject. In fact a subject is not lacking, but whereas the many predicates belong to the order of phenomena, the subject belongs to another order. Throughout the chapter of chemistry which constitutes a definition of silver, a certain ontological x unreflectingly designated by this name, silver, is present, though undisclosed, to the mind. The logically satisfactory definition of silver would be: x melts at 960.5°, boils at 2000°, etc.; we give the name of silver to the hidden essence which we circumscribe by this steadily connected set of observable regularities. Whereas the being of things is successfully penetrated by the dianoetical intellection used in philosophy of nature, it is only circumscribed by the *perinoetical* intellection of empiriological science. The intelligible element which enables empiriological knowledge to transcend empiricism is not revealed to the mind; it is neither constructed by the mind nor imposed by it upon the phenomenal matter. It is grasped by the mind inside a system of phenomenal regularities, circumscribed by this observable system and never disengaged from it. Thus the science of which Aristotle had no clear notion—although he practiced it a great deal—, the science which has for its object the phenomenal regularities themselves, is defined as possible on a realistic basis. The orderly character of the phenomena is guaranteed by the ontological *x* which is confusedly grasped together with them by the empiriological analysis. With Maritain, the science of phenomena was given for the first time a justification which owed nothing to the idealistic interpretation of the mind's activity.

It is clear that in this conception a positive science of nature can exist independently of any mathematical treatment of natural phenomena. The Kantian statement that "the amount of genuine science found in each department of natural knowledge cannot be greater than the amount of mathematics found in it" shockingly conflicts with the fact that most important developments whose scientific character can hardly be questioned seem to be by nature refractory to mathematical forms (in biology and psychology especially). Whenever the mind seizes an essence, a *ratio entis*, albeit in the blind way proper to the perinoetical intellection, a genuinely scientific treatment remains possible. Any universal and necessary form of being, however obscure may be the way it is grasped, constitutes a matter to which the mind can apply the principles of scientific thought, that is, causal and explanatory schemes. With great care Maritain pointed out that causal ideas and principles, when applied in empiriological sciences, have to be reshaped or refashioned. The concept of efficient cause, for instance, is originally an ontological concept, that is, a concept defined by reference to being; in this original condition it is not directly applicable outside the ontological order. When we go down to the empiriological level, the concept of being undergoes a transformation. Here, being no longer appears as the lighted spot of the thing under consideration, but merely as an undisclosed principle of orderliness which guarantees the steady character of the phenomenal regularities upon which light is concentrated. Causal concepts have to undergo a transmutation completely analogous to that undergone by the concept of being. This operation can make them hard to recognize, and this is how some extreme forms of positivism have been able to construe the ideal of a purely legal science which would owe nothing to causal concepts. But it is well known that the spontaneous development of positive sciences has constantly given the lie to this ideal limit of positivism.

*   *   *   *   *

Considering again the current contention that Thomism cannot account for modern epistemological developments, let us now remark that it refers especially to the mathematical aspect of modern science. Did not the Cartesian reformation consist in the substitution of a mathematical interpretation of the physical world for the Aristotelian interpretation of nature in terms of ontology?

The mathematical treatment of physical nature was not totally unknown to ancient and medieval Aristotelianism. Astronomy, optics, and acoustics are referred to in the works of Aristotle and his medieval followers as so many *mixed sciences*, whose form is mathematical and whose matter is physical. In this connection, it is necessary to correct current statements concerning the lack of explicit distinction between philosophy and positive science in ancient and medieval philosophers. Old Aristotelians failed to distinguish clearly two types of thought, corresponding to distinct degrees of abstraction within the first order, and the term *physicus* is taken by them as entirely synonymous with the term *philosophus naturalis*. In that sense it is true that up to the modern era philosophy embraced all sciences of nature. But this holds only so far as positive research assumes purely physical ways. Ancient and medieval philosophers seem to be rather keenly aware of a discrepancy between the ways proper to the *philosophia naturalis* and those proper to physico-mathematical sciences. Whereas it never occurs to them to set in opposition the *physicus* and the *philosophus naturalis*, they currently set in opposition the *philosophus physicus* and the astronomer, thus showing some realization of the non-philosophical character of the mathematical interpretation of nature.

Maritain describes the epistemological crisis which broke out at the time of Galileo and Descartes and is still so far from being settled as a *tragic misunderstanding*.[4] When the historic conflict between the Aristotelian physics and the new physics opened, both sides were equally convinced that this was a conflict between two philosophies of nature. The physico-mathematical science founded by Descartes was taken by its very founder as a philosophy of nature and the only possible one. The decadent Aristotelians with whom Descartes was confronted did not even think that the Cartesian world-picture was possibly a physico-mathematics sophisticated into an ontology. Then it happened that the Cartesian mechanism achieved the obliteration of the old distinction between the philosopher of nature (*physicus*) and the mathematical interpreter of nature (*astronomus, musicus* . . .). When we reread the great work of Newton significantly entitled *Philosophiæ Naturalis Principia Mathematica*, we realize that the Newtonian science, once considered by positivists as the archetype of positive knowledge, was far from having rid itself of ontological ambitions.

Thanks to his felicitous description of a non-philosophical approach to the physical world within the first order of abstraction, Maritain found himself in a favorable position to investigate the principles of physico-mathematical knowledge and to account for the increasingly complete autonomy which marked its latest developments. In this undertaking, Maritain had at hand two effective instruments: one was his theory on perinoetical intellection and descending analysis; the other was the conception of the mathematical object as a *preter-real* entity always affected by some *conditio rationis* and which often turns out to be a mere *ens rationis* with a foundation in the real.

It is comparatively easy to see how the law of the descending analysis which prevails in all fields of positive knowledge applies to the mathematical interpretation of nature. Whereas in the case of a non-mathematical positive science the law of descending analysis amounts to the necessity of resolving all concepts into observable data, this law, when applied to a science of physico-mathematical type, signifies the necessity of resolving all concepts into *measurable* data. Nothing makes sense for the positive scientist in general except what can be explained in terms of observations. Nothing makes sense for the physico-mathematician except what can be explained in terms of measurements. A great deal of confusion often results from the fact that the philosopher of nature and the physicist use the same terms without in most cases being aware of their referring to widely different objects. One and the same term refers to the being of things when used by the philosopher and, when used by the physicist, to the aptitude of things to be the matter of accurate measurements. No wonder that such widely different points of view give birth to statements which in appearance conflict sharply. The conflict generally vanishes as soon as we understand that identical words convey typically different concepts and refer to distinct objects. The clearest example we can think of is furnished by the recent discussion about the determination of natural phenomena. Many philosophers and scientists attribute to the so-called indeterminism of modern physics revolutionary consequences with regard to our philosophical conception of the natural and even of the human world. Yet it should be remarked that the point of reference used by the physicist in his definition of determinism is quite distinct from the point of reference used by the philosopher in the definition of a concept which bears the same name. True to the law of ascending analysis which is that of all philosophical thought, the philosopher considers that an event is determined when in some way or other it happens necessarily; necessity itself is defined as the property of that which cannot be otherwise than it is. The reference is ontological; the concept explains itself in terms of being. A concept so defined makes absolutely no sense for the physicist. Being and the possibility of being otherwise are not things which fall under his measurements. Accordingly, in order to be of any real use in physics the concept of determinism has to be reshaped so as to satisfy the following proportion: the determinism of the physicist is to the determinism of the philosopher as *the measurable* is to being. Thus we are led to realize that whereas the philosopher understands by *determined event* an event which follows from its causes in such a way that it cannot fail to happen, the physicist understands by *determined event* an event whose coordinates at the time *t* can be accurately calculated on the basis of an initial system of spatio-temporal data. The determinism of the physicist is an *empirio-metrical* determinism.

Because of the intervention of the mathematical *ens rationis* the gap is wider between philosophy of nature and physico-mathematics than between philosophy of nature and the other parts of positive science. In so far as physics is a formally mathematical science, in so far as it obeys the law which is that of its form, it participates in the indifference of mathematics to the reality of its object. This consideration accounts for the particular form taken in our times by the old conflict between science and common sense.

The congeries of current notions that we call common sense is far from being homogeneous. Maritain distinguishes in it a system of images and a rudimentary ontology. The imagery of common sense expresses mostly the laziness of uncultured intellects and their willingness to content themselves with cheap representations. No wonder that this imagery has always conflicted with science and generally with every form of rational thinking. But inasmuch as physics incorporates *entia rationis* and follows the mathematical tendency to treat indifferently *entia rationis* and *entia realia,* even the sound part of common sense, its ontology, may enter into conflict with the most sound scientific speculations. The concept of relative simultaneity, for instance, appears very shocking to common sense; common sense unhesitatingly believes that the question whether two events happen at the same time must be answered by yes or no. Ontologically considered, simultaneity is absolute. Yet the concept of relative simultaneity makes sense if referred to definite possibilities of accurate measurements; this reference is thoroughly unfamiliar to common sense. Relative simultaneity is a physico-mathematical *ens rationis* founded in the real and inescapably imposed upon the mind of the physicist by the very nature of his scientific point of view.

From this it does not follow that the constructions of the physicist should be considered as mere "hypotheses" or conventions incapable of apprehending the real in any way. Maritain would not agree with the superficial statement that the philosopher has never to worry about agreements or disagreements with the physicist, on the ground that philosophy and physics are two separate domains of thought. His epistemological pluralism is by no means absolute. Let us give an idea of the distinctions which should be made and of the phases which should be surveyed in order to appreciate the bearing of physical theories with regard to the knowledge of the real.

1. The principles previously developed make it clear that a concept may be a genuine expression of the real without pertaining to the ontological type. A description of a non-ontological character is not thereby deprived of real bearing. Real, being, knowledge are so many analogical terms. An ontological description is more real than a non-ontological one, yet a non-ontological description may well be a description of the real.

2. Even within the first order of abstraction the mind often uses fictitious constructions in its approach to the real. Yet, so long as we remain within the first order of abstraction, the realistic spirit of science is not held in check. Except for possible failures, fictions never play more than a transitional role; they are used as mere means in view of achieving a representation of the real which cannot be brought about in a more direct fashion.

3. As soon as positive science assumes a mathematical

form, something entirely novel takes place. The very nature of mathematical abstraction renders mathematical thought indifferent to the reality of its object. Consequently physico-mathematical science, in so far as it yields to the attraction of its mathematical form, tends to make no difference between *ens reale* and *ens rationis*.

4. Should this tendency prevail without check, it could be said truly that physical theories do not trace phenomena to their real causes and cannot tell anything about the real course of physical events. Such is apparently the conception of physics upheld by Pierre Duhem. For Maritain this interpretation, though not without basis, amounts to an oversimplification. As a matter of fact, the attraction exercised on physics by its mathematical form is not unchecked. If the form is mathematical, the matter remains physical and accordingly there is in the very structure of the science a counteracting tendency to stick to the real and to look for explanations by real causes. Actual science is probably a compromise between these two opposite and complementary tendencies.

*   *   *   *   *

However incomplete it may be, this exposition sufficiently shows that for Maritain the problem of the relationship between science and philosophy does not admit of any easy solution. Maritain is quite aware of the great improvements in knowledge which can be expected from the cooperation of the philosopher and the scientist; but he does not seem to believe that such a cooperation can ever work smoothly and without frictions. The vast ensemble of our knowledges of nature—philosophical, empiriological, empiriometrical—is apparently destined to present everlastingly a spectacle of restlessness, of precarious equilibrium, with sharp conflicts breaking out in times of crisis. Such a lack of harmony would be sufficiently accounted for by the psychology of the scientist and that of the philosopher. It is difficult, not to say impossible, for each of them not to be biased by his own *habitus* to the point of being unable to understand his partner. But even if a perfect philosopher were also a perfect physicist, or vice versa, there still would be within the mind provided with such *habitus* ground for conflicts between the two visions of the world. Maritain says that there is some melancholy in the realization that no complete continuity can be established among our various approaches to the natural world. It is not the least merit of his extensive and profound exploration of the most diverse fields of rational activity to have removed the optimistic illusion of a perfect harmony among the functions of the mind.

Compared with the teaching which prevailed in Thomistic textbooks thirty years ago, Maritain's philosophy of sciences appears as a tremendous novelty. Yet whoever is familiar with the physical and epistemological writings of St. Thomas will admit that no Thomist has ever written a more authentically Thomistic book than the *Degrees of Knowledge*. This great work testifies that the most living and timely expression of Thomism is not reached through eclectic combinations, but through a faithful and consistent adherence to the principles of St. Thomas. How does it happen that several philosophers, consistently faithful to St. Thomas, can do no better than voice lifeless truths, badly handicapped in the struggle against living errors? I think I understood what is wrong with these respectable thinkers when Maritain not long ago pointed out, in a letter to me, that the commentators of St. Thomas have the arduous duty of disentangling the precious stuff, bit by bit and indefatigably, from the vast amount of gangue in which it is hidden. Then, alluding to some persons whom we know well, he added: "They believe they have just to crack the shell to get the nut."

Selected Bibliography

Prepared by Donald A. Gallagher, Ph.D.

I. The Writings of Jacques Maritain on the Philosophy of Nature and the Philosophy of the Sciences
II. On Maritain's Philosophy of Nature and Philosophy of the Sciences

(For a general Maritain Bibliography, consult "Bibliography of Jacques Maritain: 1910–1942" by Ruth Byrns, *The Maritain Volume of the Thomist,* New York: Sheed and Ward, 1943; and *Jacques Maritain: Son Oeuvre Philosophique,* Paris: Desclée de Brouwer, 1948.)

Selected Bibliography

I. THE WRITINGS OF JACQUES MARITAIN ON THE PHILOSOPHY OF NATURE AND THE PHILOSOPHY OF THE SCIENCES

A. *Books*

The following books are either devoted to philosophy of nature and of science or contain important chapters and sections on these topics.

1. *Introduction générale à la philosophie. (Eléments de Philosophie, 1er fascicule.)* Paris: Téqui, 1921.
   \# English translation:
   An Introduction to Philosophy. E. I. Watkin. New York: Sheed and Ward, 1930.
2. *Théonas: ou, les entretiens d'un sage, et de deux philosophes sur diverses matières Inégalement Actuelles.* Paris: Nouvelle Librairie Nationale, 1921.
   \# English translation:
   Théonas: Conversations of a Sage. F. J. Sheed. London and New York: Sheed and Ward, 1933.
   Freedom of the Intellect, and Other Conversations with Théonas. F. J. Sheed. New York: Sheed and Ward, 1935.
3. *Antimoderne.* Paris: Éditions de la Revue des Jeunes, 1922.

185

4. *Réflexions sur L'intelligence et sur sa vie propre.* Paris: Nouvelle Librairie Nationale, 1924.

5. *Distinguer pour unir: ou, les degrés du savoir.* Paris: Desclée de Brouwer, 1932.
   # English translation:
   The Degrees of Knowledge. Bernard Wall and Margot Adamson. London: Bles, 1937; and New York: Scribner's, 1938.

6. *Le songe de Descartes, suivi de quelques essais.* Paris: Corréa, 1932.
   # English translation:
   The Dream of Descartes. Mabelle L. Andison. New York: Philosophical Library, 1944.

7. *La Philosophie de la Nature: essai critique sur ses frontières et son objet.* Paris: Téqui, 1935.

8. *Science et Sagesse, suivi d'éclaircissements sur la philosophie morale.* Paris: Labergerie, 1935.
   # English translation:
   Science and Wisdom. Bernard Wall. New York: Scribner's, 1938.

9. *Quatres essais sur l'esprit dans sa condition charnelle.* Paris: Desclée de Brouwer, 1939.

10. *Scholasticism and Politics.* Mortimer J. Adler. New York: Macmillan, 1940.

11. "The Metaphysics of Bergson" in *Ransoming the Time.* New York: Scribner's, 1941.
    French version:
    "La métaphysique de Bergson" in *De Bergson à*

Thomas D'Aquin. New York: Éditions de la Maison Française, 1944.

12. *La Philosophie Bergsonienne.* 3ᵉ édition, revue et augmentée. Paris: Téqui, 1948.

B. *Articles and Other Writings*

Many of the following articles and studies are reprinted, sometimes in revised form, in the books listed above.

1. "Le néo-vitalisme en Allemagne et le darwinisme." *Revue de Philosophie.* 6:417–441. 1910.
2. "La science moderne et la raison." *Revue de Philosophie.* 6:575–603. 1910.
3. "L'évolutionnisme de M. Bergson." *Revue de Philosophie.* 19:467 ff. 1911.
4. "A propos de la révolution cartésienne: philosophie scholastique et physique mathématique." *Revue Thomiste.* N.S. 1: 158–181. 1918.
5. Preface to: Driesch, Hans. *La philosophie de l'organisme.* . . . Traduction de M. Kollmann. Paris, 1921.
6. "Philosophie et science expérimentale." *Revue de Philosophie.* 33: 342–378. 1926.
7. "Philosophie et science expérimentale." In Cahiers de Philosophie de la Nature, *Mélanges* (1re Série), Paris: Vrin, 1929, pp. 159–211.
8. "De la notion de la philosophie de la nature." *Philosophia Perennis.* 2: 819–828. Muenchen: Habbel, 1930.

9. Notes to: André, H.; Buytendijk, F. J. J.; Dwelshauvers, G.; Manquat, M. *Vues sur la psychologie animale.* Paris: Librairie Philosophique J. Vrin, 1930.

10. "Science et philosophie d'après les principes du réalisme critique." *Revue Thomiste.* 36: 1–46. 1931.

11. "La philosophie de la nature: philosophie et sciences." *La Vie Intellectuelle.* 31: 228–259. 1934.

12. "Philosophie de la nature et sciences expérimentales." *Acta Pont. Academiae Romanae S. Thomae Aq. et Religionis Catholicae.* 1: 77–93. 1934.

13. "Notes pour un programme d'enseignement de la philosophie de la nature et d'enseignement des sciences dans une Faculté de Philosophie." *Bollettino Filosofico.* 1, n. 2: 15–31. 1935.

14. "Philosophie de l'organisme. Notes sur la fonction de nutrition." *Revue Thomiste.* 43:263–275. 1937.

15. "Science et philosophie." *Acta Pontificae Academiae Romanae.* N.S. 3:250–271. 1937.

16. "Good Will in Science." *The New York Times.* Section IV, p. 7, August 4, 1940.

17. Foreword to: Adler, M. J., *The Problem of Species.* New York: Sheed and Ward, 1940.

18. "Concerning a Critical Review." *The Thomist.* 3: 45–53. 1941.

19. "The Conflict of Methods at the End of the Middle Ages." *The Thomist.* 3: 527–538. 1941.

20. "Science, Philosophy and Faith." In *Science, Philosophy and Religion.* New York: Conference on

Science, Philosophy and Religion, 1941, pp. 162–
183.

21. "Science and Wisdom." In *Science and Man*,
twenty-four original essays edited by Anshen, R. N.
New York: Harcourt Brace, 1942.

II. ON MARITAIN'S PHILOSOPHY OF NATURE
AND PHILOSOPHY OF THE SCIENCES

1. Renoirte, F. "La philosophie des sciences selon M.
Maritain." *Revue Néo-Scholastique*. 35: 96–106.
1933.
2. Regis, L. M. "La philosophie de la nature: quelques
'apories'." In 1er Cahier de *Philosophie* (Ottawa).
pp. 127–156. 1936.
3. Phelan, G. B. *Jacques Maritain*. New York: Sheed
and Ward, 1937.
4. Simon, Y. "Maritain's Philosophy of the Sciences."
In *The Maritain Volume of the Thomist (Thomist
V)*. New York: Sheed and Ward, 1943, pp. 85–102.
5. Lima, Alcen Amoroso. (Tristao de Athayde), "A
filosofia sintetica de Maritain." in *Jacques Maritain*.
(*A Ordem*, Brazilian Review), 1946.
6. Leroy, Marie-Vincent. "Le savoir spéculatif." In
*Jacques Maritain, Son Oeuvre Philosophique. (Revue
Thomiste*, t. XLVIII). Paris: Desclée de Brouwer,
1948, pp. 236–339.
7. Gerardus, Brother. "Jacques Maritain Doctus An-

gelico." *Journal of Arts and Letters.* 1 (nos. 1–3). 1949.

8. Klubertanz, G. P. "What is the Philosophy of Human Nature?" In *Notes on the Philosophy of Human Nature* (chap. XIV). Saint Louis: Saint Louis University, 1949.

9. Gallagher, D. A. "Contemporary Thomism." In *A History of Philosophical Systems.* New York: Philosophical Library, 1950. (Contains a general summary of Maritain's philosophy.)

Footnotes

PREFACE

1. *Sept Leçons sur l'Être*, Paris, Téqui, 1934, p. iii. *Preface to Metaphysics (Seven Lectures on Being)*, New York, Sheed and Ward, 1937; preface not included.

CHAPTER I

1. *Les Degrés du Savoir* (Paris, Desclée de Brouwer, 1932), ch. II, p. 43 f.; ch. IV, p. 265 f. *The Degrees of Knowledge* (London, Centenary Press, 1937); Ch. I, pp. 27-85; Ch. III, pp. 165-247.
2. Cf. St. Thomas Aquinas, *In Periherm.*, lib. I, cap. vii, lect. 10, n. 9 (capital text for the theory of the universal).
3. St. Thomas Aquinas, *In Meta.*, lib. iv, lect. 4. Cf. also *ibid. Procemium.*
4. Aristotle, *Physics* III, 1, 200b 12.
5. "Quaedam igitur sunt speculabilium quae dependent a materia *secundum esse,* quia non nisi in materia esse possunt; et haec distinguuntur, quia dependent quaedam a materia *secundum esse et intellectum,* sicut illa in quorum definitione ponitur materia sensibilis: unde sine materia sensibili intelligi non possunt, ut in definitione hominis oportet accipere carnem et ossa, et de his est physica sive scientia naturalis. Quaedam vera quamvis dependeant a materia *secundum esse, non tamen secundum intellectum,* quia in eorum definitionibus non ponitur materia sensibilis, ut linea et numerus, et de his est mathematica. Quaedam vero sunt speculabilia quae *non dependent a materia secundum esse;* quia sine materia esse possunt, sive nunquam sint in materia, sicut Deus et angelus, sive in quibusdam sint in materia et in quibusdam non, ut substantia, qualitas et actus,

unum et multa, et hujusmodi, de quibus omnibus est theologia, id est divina scientia, quia praecipuum cognitorum in ea est Deus. Alio nomine dicitur metaphysica, id est transphysica, quia post physica discenda occurrit nobis, quibus ex sensibilibus competit in insensibilia devenire. Dicitur etiam philosophia prima, in quantum scientiae aliae ab ea principia sua accipientes eam sequuntur." St. Thomas, *in lib Boet. de Trinit.*, q. 5, a I resp.

Cf. *ibid.*, a. 3, resp. In this article St. Thomas gives a detailed explanation of the nature and the differences between the three degrees of abstraction, particularly how mathematical abstraction "respondet unioni formae et materiae, vel accidentis et subjecti; et haec est abstractio formae (sc. quantitatis) a materia sensibili," and how the first degree of abstraction "respondet unioni totius et partis: et est abstractio totius, in qua consideratur absolute natura aliqua secundum suam rationem essentialem, ab omnibus partibus, quae non sunt partes speciei, sed sunt partes accidentales."

6. St. Thomas Aquinas, *in lib. Boeth. de Trinit.* q. 5, a. 3.
7. Cajetan, Prooemium *in De Ente et Essentia*, q. 1, circa primum.
8. Cajetan, *loc. cit.* This is also the case, proportionately, for the objects of mathematics. Mathematics too, is as form to matter in relation to the objects of the physicist.
9. *Loc. cit.*
10. St. Thomas, *in lib. Boeth. de Trin.* q. VI, art. 2, resp.
11. *Ibid, loc. cit.*
12. St. Thomas, *In Meta.* VI, lect. 1.
13. Étienne Gilson, *Études sur le rôle de la pensée mediévale dans la formation du système cartesien*, Paris, Vrin, 1930; Ch. III Météores cartésiens et météores scolastiques; pp. 102–137.

CHAPTER II

1. See below, page 73 and following.
2. Émile Meyerson, *De l'explication dans les sciences* (Paris, Payot 1921); vol. I, p. 6.

3. Auguste Comte, *Cours de philosophie positive* (4me édit., Paris, 1887), vol. II, p. 312. Quoted by Meyerson, *op. cit.*, vol. I, p. 6.

4. Goethe, *Faust*, Part 1, sc. 4, l. 1936–1939.

5. Goethe, *Faust*, Pt. 1, sc. 1, line 417.

6. Goethe, *op cit.*, Pt. 1, sc. 1, l. 377.

7. The "subjective synthesis" of Auguste Comte has to do with practical philosophy.

8. Meyerson, *op. cit.*, vol. I, ch. 1.

9. Antoine A. Cournot, Traité de l'enchainement (Paris, Hachette, 1861), vol. I, p. 264. Quoted by Meyerson, *op. cit.* I, p, 26.

10. Meyerson, *op. cit.* I, p. 31.

11. *Ibid.*, p. 31.

12. *Ibid.* I, ch. 2, p. 36.

13. *Ibid.* I, ch. 2, p. 39.

14. Arthur S. Eddington, *The Nature of the Physical World* (Cambridge Univ. Press, 1933), p. 209.

15. Quoted by Meyerson, *op. cit.* I, p. 45.

16. Meyerson, *op. cit.*, I, p. 180.

17. Confer e.g. *De l'explication dans les sciences*, ch. 6, p. 181; ch. 9, p. 311 f.

18. Gaston Bachelard, *Le nouvel esprit scientifique* (Paris, Alcan 1934).

19. See below, page 105. *Les degrés du savoir*, ch. IV, p. 273 f. *The Degrees of Knowledge*, ch. III, p. 168 f.

CHAPTER III

1. John of St. Thomas, *Cursus Philosophicus*, Log. II, q. xxvii, art. 1. "Quia in rebus materialibus, quae redduntur intelligibiles et immateriales per segregationem a materia et conditionibus materialibus, ipsa abstractio est quasi motus quidam, in quo consideratur terminus a quo et terminus ad quem, formaliter quidem in ipso actu abstractionis, fundamentaliter vero et objective in ipso objecto abstrahibili. Ex parte termini a quo habet derelictionem materiae, quae triplex est,

ut supra diximus, et sic constituitur triplex genus abstractionis. Ex parte autem termini ad quem est diversus gradus immaterialitatis seu diversus modus spiritualitatis, quem acquirere potest res sic abstracta. Et hoc vocat D. Thomas 1. *Poster.* lect. 41—in unoquoque genere scibilitatis distingui diversas species secundum diversos modos cognoscibilitatis. —Quare non solum sumitur ratio formalis et specifica scientiarum ex recessu a materia, sed ex accessu ad determinatum gradum immaterialitatis, quo objectum aliquod determinate deputatur et redditur intelligibile; sicut etiam in angelis diversa species non solum sumitur ex recessu a corporeitate, sed ex accessu ad determinatum modum habendi spiritualitatem et immaterialitatem et ad actum purum, ut S. Thomas advertit 6. cap. de Ente et Essentia, circa finem. Unde in Mathematicis invenimus, quod licet in communi abstrahant a materia sensibili, tamen quia diversus modus immaterialitatis attingitur in quantitate continua quam discreta, discreta enim minus concernit materialitatem, quia minus dependet a loco et tempore quam continua, quae copulat partes suas in loco, ideo duplex scientia constituitur, Geometria et Arithmetica. Et similiter Philosophia et Medicina duplex scientia est, quia licet utraque abstrahat a materia singulari, tamen magis concernit materiam corpus ut sanandum quam corpus mobile ut sic. . . .

"Quare cum specificatio atoma scientiarum sit ultima ratio scibilitatis, quae non est amplius divisibilis, oportet, quod si ratio formalis scibilitatis sumitur ex immaterialitate, ultima et specifica sumatur determinate ex termino ad quem talis abstractionis, in quo ultimo sistit et determinatur abstractio. Ergo non ex sola segregatione a materia, prout consideratur terminus a quo abstractionis, sed in ultima determinatione immaterialitatis specfica et determinata ratio scibilitatis consistet."

2. *Ibid., sub fine.*

3. John of St. Thomas, *Philosophia Naturalis* I, q. 1, art. 2. Also, in the same article: "Tunc autem diversa abstractio fundat diversam speciem, et modum illuminandi, quando

oritur ex diversis principiis: ex principiis enim sumitur illu-
minatio conclusionum, ut late tractavimus in libris Posteri-
orum, q. xxvii. Cum autem principia quibus passiones pro-
bantur de subjecto, sunt definitiones, ideo *ad diversum
modum definiendi* reducitur diversa species sciendi, et mani-
festandi res scitas; non enim sufficit alias et alias res definire
et tractare, sed aliter atque aliter: nam plures res definire et
de pluribus quidditatibus agere, etiam in una scientia con-
tingit, quatenus omnia illa sub uno modo definiuntur, ut
citato loco Logicae ex D. Thoma Metaphys. lect. 1 et pluribus aliis locis ostendimus. Omnes autem definitiones, quae
traduntur tam in octo libris Physicorum quam in libris de
Generatione et aliis, sub eadem abstractione et formalitate
procedunt."

4. Cf. St. Thomas Aquinas, *Comment. in De Sensu et Sensato*,
   lect. 1.

5. John of St. Thomas, *Cursus Philosophicus*, Log. II, q. xxvii,
   art. 1; cited above, note 1, p. 193.

6. Aristotle, *Physics*, bk. VIII, 5, 256 ff.

7. Cf. John of St. Thomas, *Cursus Philos.*, Log. II, q. xxvi,
   art. 1.

8. Chapter II, n° 4, page 53.

9. This rendering of *"appel d'intelligibilité"* was chosen, despite the emotional connotations of the English "appeal,"
   because of the need for a compact expression. It is admit-
   tedly not completely satisfactory. Mr. Maritain's subtle formula suggests, by analogy with *"appel d'air,"* that in knowing a thing one stands in a 'draft of intelligibility' coming
   from that thing as one might stand in a draft of air coming
   from a window, for example. This 'draft' is, as it were, the
   thing's intelligible breath, the intelligible voice with which
   it speaks to us or *summons* us to know it. Tr.

10. As John of St. Thomas remarks (*Cursus Phil.*, Philos. Nat.
    IV, q. 2, a. 3, *secunda difficultas*) the objective light may
    be taken in two ways: 1. from the side of the knowing power
    or *habitus*; 2. from the side of the object. "Ratio formalis
    sub qua sumitur dupliciter, uno modo ut tenet se ex parte

potentiae seu habitus, et sic est ipsa ultima ratio virtutis qua determinatur et proportionatur erga tale objectum. Alio modo sumitur ex parte ipsius objecti, et sic est ultima formalitas proportionans et coaptans objectum potentiae vel actui . . ." We are taking it in the second way here, considering it only *ex parte ipsius objecti, ut ultima formalitas proportionans et coaptans objectum habitui.*

11. Pierre Duhem, La Théorie physique, Paris (Chevalier & Rivière, 1906), p. 265.

12. P. Descoqs, *Essai critique sur l'hylémorphisme*, Paris (Beauchesne), 1925.

13. Hans Driesch, *La philosophie de l'organisme*, Paris, Rivière 1921.

14. Hans André, *Urbild und Ursache in der Biologie*. Munich and Berlin, Oldenbourg 1931.

15. *Cahiers de philosophie de la nature*, edited by Rémy Collin and R. Dalbiez (Paris, Vrin, 1927, vol. 1—1930, vol. 4).

16. *Les dégrés du savoir*, ch. IV, p. 368 ff.; *The Degrees of Knowledge*, ch. III, p. 228 ff.

17. Aristotle, *Physics* IV, 13, 222b, 19–20; IV, 12, 221b 1–2.

CHAPTER IV

1. Main writings of Maritain concerning the philosophy of sciences: *Réflexions sur l'intelligence*, Paris, 1924, Ch. 6 and 7; *Distinguer pour unir ou les Dégrés du savoir*, Paris, 1932 (English translation, *The Degrees of Knowledge*, Scribner's, New York, 1938); *La Philosophie de la nature*, Paris, 1935; *Science and Wisdom*, Scribner's, New York, 1940; *Scholasticism and Politics*, Macmillan, New York, 1940, Ch. 2.

2. It goes without saying that in this sketch we content ourself with pointing out major features of the systems under consideration.

3. I abstract from the question whether an empiriological species like silver coincides with an ontological species, or is merely a sub-determination of a broader ontological species.

4. See *The Conflict of Methods at the end of the Middle Ages*, THE THOMIST, Oct. 1941.

Index to Proper Names

André, Hans, 71, 108, 151
Aristotle, 2, 7–13, 27, 31, 32, 33, 37, 38, 41, 84, 95, 150, 155, 160, 161, 168, 169, 171, 173

Bachelard, Gaston, 69, 70
Bergson, 58, 59, 121, 152
Biran, Maine de, 56
Blondel, 22
Bohr, 149, 151
Brentano, 71
Broglie, Louis de, 2, 151
Brunschvicg, 22, 57, 70, 121, 152

Cajetan, 11, 16, 17, 20, 21, 23, 118, 120, 123, 125–129
Collin, Rémy, 151
Comte, Auguste, 51, 52, 54, 55, 57
Conrad-Martius, Mme. Hedwig, 84
Cournot, Antoine, 63
Cousin, Victor, 56, 159

da Vinci, Leonardo, 37
Descartes, 25, 35, 36, 43–45, 59, 82, 84, 88, 90, 157, 174

Descoqs, Father, 145, 146
Dirac, 151
Driesch, 1, 58, 151
Duhem, Pierre, 60–62, 64, 69, 142, 180

Eddington, Arthur, 65
Einstein, 2, 77, 151

Foucault, 57
Friedmann, 84

Galileo, 36, 95, 157, 174
Gilson, Étienne, 35
Goethe, 54, 153
Goudin, 34
Gredt, Father, 71

Heisenberg, 78, 151
Heraclitus, 4, 5, 11, 12, 167
Husserl, 71, 121

John of St. Thomas, 34, 91, 102, 103, 161, 162

Kant, 37, 46–48, 51, 159, 168, 169
Kelvin, Lord, 65

197

Leibniz, 59
Littré, 55
Lorenz, 151

Mach, 52
Maritain, Jacques, 59, 160–
    162, 170, 172, 174, 175,
    178–182
Melissus, 119
Meyerson, Émile, 48, 52, 62–
    70, 93, 152

Newton, 36, 175

Parmenides, 5, 11, 12, 119,
    167
Planck, 2, 151
Plato, 4–9, 11, 12, 167

Plessner, 84
Poincaré, Henri, 151
Pythagoras, 57

Russell, Bertrand, 152

Scheler, Max, 71, 152
Schrödinger, 149
Socrates, 12

Thomas Aquinas, St., 14, 16–
    18, 25, 26, 32, 38, 91, 104,
    155, 157–161, 181, 182

Victoria, Queen, 65

Whitehead, 152
Wolff, 1, 32, 159, 160